Modern Astrodynamics

Second Edition, 2010

William E. Wiesel
Professor of Astronautical Engineering
Air Force Institute of Technology
Wright-Patterson AFB, OH.

Published by *CreateSpace*
https://www.CreateSpace.com/3479062

ISBN 978-145378-1470

First Edition: 2003.
Second Edition 2010.

Wiesel, William E.

Modern Astrodynamics / William E. Wiesel. –2nd ed.
p. cm.
Includes index
TL 1050.W 2004
ISBN 978-145378-1470
1. Astrodynamics I. Title

Preface

This book grew out of a series of courses given at the Air Force Institute of Technology during the last two decades. It is an attempt to blend the rich history of perturbation theory in orbital mechanics, the legacy of three centuries of development, with the modern theories and methods being developed around the capabilities of the digital computer. The field of astrodynamics is being given a new immediacy as the computer enables us to see deeper into the actual structure of the solutions to a problem, while the older methods have lost none of their insight. Throughout, I have attempted to develop both the older "classical" approach to perturbation theory with the more modern canonical techniques, and I end with two chapters that are examples of the marriage of the new computer techniques with the older perturbation methods.

The book presumes facility with Lagrangian and Hamiltonian dynamics, and some slight familiarity with Hamilton's principle and the methods of the calculus of variations. This material is usually covered in a course in analytical mechanics, or a second course in dynamics.

W. Wiesel

Contents

Chapter 1

Modern Numerical Methods

1.1 Introduction

The name special perturbations is the usual term in orbital mechanics for numerical integration of the equations of motion. The term refers to the fact that numerical integration produces results that are special to that one particular orbit. In contrast, the technique of general perturbations will produce results which could be applied to many different orbiting objects, simply by changing the initial conditions which appear in the series approximations. This is not an option when an orbit is numerically integrated, since if you want results for a new orbit, you must numerically integrate that new orbit.

For most of the history of this discipline, special perturbations was not really a viable option. The labor required for a complete integration of the equations of motion far exceeded the effort that could be reasonably expended on one orbit. Also, the technique of general perturbations could give results for many different objects, and furthermore could produce a prediction at just the one instant in time for which data was available. There was no need to numerically integrate through a long time interval to get an object's position at one new time. Finally, most of the natural objects in the solar system have orbits which do not force the use of special perturbations: small eccentricities and inclinations lead to rapidly convergent series expansions, while the lack of close approaches ensures that perturbations remain small.

For general perturbations assumes that the object only experiences small deviations from two body motion. There are cases for which this is not true: an interplanetary spacecraft is one good example, comets which suffer close approaches to Jupiter and have their orbits radically altered are another. These cases require the use of numerical integration. Another case is very high eccentricity orbits, where the usual expansion in the eccentricity does not converge rapidly, if it converges at all. In fact, the first important use of special perturbations was due to Cowell and Cromellin, who numerically integrated the orbit of Halley's comet from the 1840's appearance forward to the 1910 return. According to legend they began *years* before the predicted reappearance of the comet, and only beat it by a few months!

The most obvious technique, direct numerical integration of the equations of motion in rectangular inertial coordinates, is still referred to as Cowell's Method in celestial mechanics. There are more techniques available today to handle numerical problems in celestial mechanics, not to mention the staggering increase in computational capabilities since the time of Cowell. His problem would take only a few moments on an IBM type personal computer, not to mention using a "real" computer on this problem.

But in the last decades numerical integration has become much more than a tool for just propagating a single orbit. We will see in this chapter that it offers a way to peer into the inner workings of the most nonlinear system, and permits asking all manner of new and interesting questions. Also, numerical integration makes it possible to easily add forces that are not easily treated by classical perturbation theory. This is almost certainly the future of orbital mechanics.

1.2 Cowell, Encke, Lagrange, et al

Once the decision has been made to resort to numerical integration of the equations of motion, there are still many forms of the equations of motion which could be employed. Generically, any integration in physical coordinates, especially rectangular coordinates, is termed Cowell's method. We might integrate the set

$$\dot{\mathbf{x}} = \frac{d}{dt}\begin{pmatrix} \mathbf{r} \\ \mathbf{v} \end{pmatrix} = \left\{ \begin{array}{l} \mathbf{v} \\ -\mu\mathbf{r}/r^3 + \mathbf{a}_p \end{array} \right. \tag{1.1}$$

where $\mathbf{a}_p$ is the perturbing acceleration. The first thing to note about these equations is that many of our significant figures in the integration are simply going to reproduce the dominant two body part of the solution. This is especially true if the perturbations are very small, in which case the result might not contain anything *but* the two body solution.

On the other hand, a Cowell's method is easy to implement, and might well be all that is required. Direct numerical integration of the physical coordinates carrying ten significant figures (easily done) corresponds to an accuracy of $\pm \approx 3$ *centimeters* for an earth – moon trajectory (more than is needed) or $\pm \approx 1$ *kilometer* in the solar system (possibly adequate). This assumes that you still have ten digit accuracy at the end of the integration. The longer the integration period, the harder it is to retain accuracy in a numerical integration. The small errors made by using finite precision arithmetic, roundoff and truncation error, tend to build up as the square root of the number of calculations performed. Over very long periods, then, numerical integration is highly suspect; while for short periods it can easily exceed any rational accuracy requirement. Interplanetary and lunar trajectories generally require less than one revolution, and since they keep changing their allegiance to different "primary" bodies, they are prime candidates for direct integration in physical coordinates.

However, in bygone days when the only computational aids were slide rules (3 digits), logarithm tables (5 – 6 digits), and hand cranked mechanical calculators[1] a ten digit numerical integration was virtually out of the question. Special perturbations was only viable if some way could be found to reduce the labor required. In the old days, this meant reducing the number of significant figures required in the integration. Using our knowledge of the solution to the two body problem seems natural, so, following Encke, introduce a reference orbit obeying the two body equations of motion

$$\ddot{\vec{\rho}} = -\mu \frac{\vec{\rho}}{\rho^3} \tag{1.2}$$

Now, write $\mathbf{r} = \vec{\rho} + \delta\mathbf{r}$, and notice that

$$\delta\ddot{\mathbf{r}} = \ddot{\mathbf{r}} - \ddot{\vec{\rho}} = -\mu\left(\frac{\mathbf{r}}{r^3} - \frac{\vec{\rho}}{\rho^3}\right) + \mathbf{a}_p \tag{1.3}$$

Now, the bracketed term should be small, but we are forced to calculate it as the small difference of two large and nearly equal numbers. This requires a large number of significant figures, which is exactly what we are trying to avoid.

Encke's method of avoiding this problem was somewhat different, but in modern terms we can eliminate this problem with a simple series expansion. Write

$$\begin{aligned} r^{-3} &= |\vec{\rho} + \delta\mathbf{r}|^{-3} = \left\{\rho^2 + 2\vec{\rho}\cdot\delta\mathbf{r} + \delta r^2\right\}^{-3/2} \\ &\approx \rho^{-3} - \frac{3}{2}\frac{1}{\rho^5}\left(2\vec{\rho}\cdot\delta\mathbf{r} + \delta r^2\right) + \ldots. \end{aligned} \tag{1.4}$$

Inserting this result into (1.3), and truncating after the first order in $\delta\mathbf{r}$, we find

$$\delta\ddot{\mathbf{r}} = -\mu\left\{\frac{\delta\mathbf{r}}{\rho^3} - 3\vec{\rho}\cdot\delta\mathbf{r}\frac{\vec{\rho}}{\rho^5}\right\} + \mathbf{a}_p + \mathcal{O}(\delta r^2) \tag{1.5}$$

So long as second order terms in $\delta\mathbf{r}$ remain negligible, equation (1.5) enables us to integrate only the perturbation. If $\delta\mathbf{r}$ grows inconveniently large, we can always declare the current trajectory to be the starting point of a new two body reference solution, and $\delta\mathbf{r}$ is small again. All of our precious significant figures go into integrating the perturbations, and not the two body reference solution.

However, as long as we are going to do this, the natural question should be: Why not just integrate the Lagrange Planetary Equations? In the old days

[1] That's where the expression 'cranking out the numbers' comes from.

this was not done, because the Lagrange Planetary Equations contain many computationally expensive square roots and trigonometric functions. Now, any of the forms of the perturbation equations of motion discussed later in Chapter 2 are suitable for numerical integration. Some are better adapted to some problems than are others, however. Using the raw planetary equations to integrate a geosynchronous, equatorial orbit would enable you to discover first hand just what the $e \Rightarrow 0$, $i \Rightarrow 0$ singularities mean. The Delaunay equinoctal elements are a far better choice in this case.

Although we will often center the discussion of this chapter on numerical integration in physical coordinates, the methods we will develop are equally applicable to integration of any of the perturbation equations of motion.

1.3 Numerical Integrators

Almost any technique for obtaining equations of motion will eventually lead to a system of equations

$$\dot{\mathbf{x}} = \mathbf{f}(\mathbf{x}, t) \tag{1.6}$$

in first order form. Both Hamilton's equations and the Lagrange Planetary equations are in this form, and either can be numerically integrated. Approaches which lead to second order equations (Newton's second law, Lagrange's method) can easily be reformulated as first order equations. Most numerical integration packages assume that the system has been brought into the form (1.6), and they require that the user supply a main program which sets up initial conditions, controls input and output, and (often) sets the timestep. In addition, the user will be required to supply a subroutine containing the actual equations of motion (1.6). That is, a subroutine which given the state vector $\mathbf{x}$ and the time will calculate the right hand sides of the equations of motion.

The variables of the problem have been grouped into a column vector $\mathbf{x}$, which might be a vector of canonical coordinates and momenta, or the vector of the six classical elements. In the former case, the state vector is just the phase space position vector of the system. Notice that the units do not need to be the same on each element of the state vector. However, it is safer numerically if all of the state vector elements have about the same characteristic order of magnitude. This is one good reason for using nondimensional variables in the formulation of a problem. (If you insist upon using distance units of millimeters, and velocity units of lightyears per nanosecond, you deserve the numerical problems you will have!)

Numerical integrators fall into several classes. Perhaps the simplest are extrapolators. In a second order system, we can use the equations of motion to evaluate the particle acceleration $\mathbf{a}$ at some time t. At some slightly later time, assuming that the acceleration is nearly constant, we might construct the position and velocity vectors as

$$\mathbf{r}(t + \Delta t) \approx \mathbf{r}(t) + \mathbf{v}(t)\Delta t + \frac{1}{2}\mathbf{a}\Delta t^2 \tag{1.7}$$

$$\mathbf{v}(t + \Delta t) \approx \mathbf{v}(t) + \mathbf{a}\Delta t$$

As $\Delta t \to 0$, two things happen: the approximation in (1.7) converges to reality, and our computer execution time diverges to infinity. Higher order extrapolators can be constructed by evaluating the time derivatives of the equations of motion themselves, and retaining more terms in the Taylor's series (1.7).

However, after we have performed several steps of the above process, we will have several previous values of the state vector $\mathbf{x}$ and its rate of change $\mathbf{f}$ lying around. These can be used as data points through which we may run polynomials in time, in which case the method is said to be a predictor method.

Both extrapolators and predictors step their way into the future using data from the current instant and the immediate past. However, once a new point at $t + \Delta t$ is available, the system equations of motion can be evaluated at that point. This is new information which was not available to us before we made the prediction, and the question arises if this information can be used to improve the quality of the new state vector. If a higher order polynomial is run through the previous data points, *and* the new equations of motion evaluated at $t + \Delta t$, you have what is termed a predictor – corrector method. For example, a straight line predictor (the trapezoidal rule) might be followed with a quadratic corrector step (Simpson's rule). Predictor – corrector methods can be very fast, and need not suffer from the divergence characteristic of extrapolators used alone.

However, there is another issue here, as first recognized by the numerical analyst Hamming. The simple predictor – corrector method mentioned above cannot successfully integrate $dx/dt = 0$ without diverging exponentially! Hamming realized that a predictor corrector method is a set of linear finite difference equations, whose forcing function is the actual system to be integrated. Just as with linear differential equations, the solution consists of a homogeneous part and a particular part, and in this case we are only interested in the particular part. If the homogeneous system (the integrator algorithm alone) is itself unstable, then eventually the actual solution function is buried under the exponential divergence of the unstable homogeneous part. It is not sufficient to start the homogeneous part at zero, since roundoff and truncation error will eventually excite the homogeneous solution. So, it is necessary that the numerical integration algorithm be numerically stable, a concept which has nothing to do with dynamical stability (except that the integrator itself is a dynamical system!). This is sufficient reason to leave the construction of such methods to specialists.

One drawback of predictor – correctors is that they need several points to begin, not just a set of initial conditions. For example, a fourth order predictor corrector needs four initial points. Since you have only one point (the initial conditions), this means that predictor correctors are not self – starting. Rather, they need an initialization method to get the other points before they can be called into operation. This makes changing timestep rather expensive with a predictor – corrector.

Numerical integrators are rated on their order...the highest order polynomial system that they will integrate exactly (or to within roundoff error). For example, a fourth order algorithm means that its error falls off as

$$\epsilon \propto \mathcal{O}(\Delta t^5) \tag{1.8}$$

as the timestep Δt is shortened for nonlinear systems. There is a tradeoff between high order integrators, which may be able to use larger timesteps, and the complexity of a high order method. "Large" timesteps can also be dangerous if they cause the integrator to blithely jump over a sudden, violent change in the system.

1.4 The Equations of Variation

If the system equations of motion

$$\dot{\mathbf{x}} = \mathbf{f}(\mathbf{x}, t) \tag{1.9}$$

are numerically integrated, we are still faced with the problem of knowing *nothing* about any other orbit. Usually, we are not sure of the exact orbit we want, or the actual orbit the satellite is following. We expect that the real orbit we are interested in is nearby, but we probably (read: absolutely certainly) do not have the real orbit. Rather than trying new sets of initial conditions at random in the hope that things will improve, there is a standard way to obtain information about all nearby trajectories at the same time that the initial orbit is integrated.

Suppose our set of initial conditions $\mathbf{x}_o(t_o)$ leads to the trajectory $\mathbf{x}_o(t)$ during the numerical integration. Let us write a general nearby trajectory as

$$\mathbf{x}(t) = \mathbf{x}_o(t) + \delta\mathbf{x}(t) \tag{1.10}$$

and substitute this into the equations of motion (1.9). Expanding the equations of motion in a Taylor's series about $\delta\mathbf{x} = 0$, we get

$$\begin{aligned} \dot{\mathbf{x}} &= \dot{\mathbf{x}}_o + \delta\dot{\mathbf{x}} = \mathbf{f}(\mathbf{x}_o + \delta\mathbf{x}, t) \\ &\approx \mathbf{f}(\mathbf{x}_o, t) + \left.\frac{\partial \mathbf{f}}{\partial \mathbf{x}}\right|_{\mathbf{x}_o} \delta\mathbf{x} \end{aligned} \tag{1.11}$$

Since $\mathbf{x}_o$ is itself a solution to the equations of motion, we may cancel equation (1.9) from both sides of (1.11) to obtain

$$\delta\dot{\mathbf{x}} = \left.\frac{\partial \mathbf{f}}{\partial \mathbf{x}}\right|_{\mathbf{x}_o} \delta\mathbf{x} \tag{1.12}$$

The partial derivative of a vector $\mathbf{f}$ with respect to a vector $\mathbf{x}$ is a square matrix, which we will abbreviate as $A(t) = \partial\mathbf{f}/\partial\mathbf{x}$. It is the matrix of partial derivatives of the equations of motion with respect to the variables in the problem, and after evaluation on the nominal trajectory $\mathbf{x}_o$, it is a function of time alone.

Now, the system (1.12) is termed the *equations of variation*, and they form a linear, time varying set of differential equations. However, since they are a linear set of equations, the addition of any two solution functions is still a solution, and the general solution can be constructed from a fundamental set of solutions. Define N independent solutions $\vec{\phi}_i(t)$ to (1.12) to have initial conditions $\phi_{ij}(t_o) = \delta_{ij}$, where δ_{ij} is Kroenecker's delta. Since only element $i = j$ of $\vec{\phi}_i$ is nonzero at $t = t_o$, a solution satisfying a general set of initial conditions can be written

$$\delta\mathbf{x}(t) = \sum_{j=1}^{N} \vec{\phi}_j(t)\delta x_j(t_o) \tag{1.13}$$

Now, a function whose initial conditions are constructed as (1.13) at the initial time will still be a solution at any other time t. Also, the function (1.13) has the correct initial conditions for any nearby orbit which we might wish to specify. So, (1.13) is the general solution to (1.12).

It is much more common to replace (1.13) with a matrix formulation. Define the matrix $\Phi(t, t_o)$ to be the square matrix whose columns are the $\vec{\phi}_i$. Then, Φ satisfies the differential equation

$$\dot{\Phi}(t, t_o) = A(t)\Phi(t, t_o) \tag{1.14}$$

with initial conditions

$$\Phi(t_o, t_o) = I \tag{1.15}$$

the identity matrix, and (1.13) can be rewritten as

$$\delta\mathbf{x}(t) = \Phi(t, t_o)\delta\mathbf{x}(t_o) \tag{1.16}$$

which is the general solution to (1.12). So, by numerically integrating the equations of variation (1.14) in parallel with the equations of motion (1.9), we can learn everything there is to know about trajectories "close" to the nominal trajectory we are integrating.

The matrix Φ is called the *state transition matrix* by deluded people who really do think that the universe is a linear system. It might be better termed the *differential* state transition matrix, because that is what it does in a nonlinear system. The state transition matrix obeys the identities

$$\Phi(t_2, t_0) = \Phi(t_2, t_1)\Phi(t_1, t_0) \tag{1.17}$$

$$\Phi(t_0, t_1) = \Phi^{-1}(t_1, t_0) \tag{1.18}$$

whose proofs are quite simple.

Numerical integration of the equations of variation is expensive. For a system with N equations of motion, there are N^2 equations of variation (1.14), requiring the simultaneous numerical integration of $N^2 + N$ first order differential equations. For a single particle with three degrees of freedom, this is 42 coupled ordinary differential equations. There is an alternate method available to obtain Φ in the case where the actual solution to the system (1.9) is in hand. It should be obvious from (1.16) that the state transition matrix is the derivative of the state solution at time t with respect to the initial conditions at time t_o. If not, let the actual solution function be $\mathbf{x}(t) = \mathbf{S}(\mathbf{x}(t_o), t)$, a function of time and the initial conditions. Expanding this about the nominal initial conditions, we have

$$\mathbf{x}_o(t) + \delta\mathbf{x}(t) \approx \mathbf{S}(\mathbf{x}_o(t_o), t) + \left.\frac{\partial \mathbf{S}}{\partial \mathbf{x}}\right|_{\mathbf{x}_o} \delta\mathbf{x}(t_o) \tag{1.19}$$

Performing the obvious cancellation and comparing to (1.16), we have

$$\Phi(t, t_o) = \left.\frac{\partial \mathbf{S}}{\partial \mathbf{x}(t_o)}\right|_{\mathbf{x}_o} = \frac{\partial \mathbf{x}(t)}{\partial \mathbf{x}(t_o)} \tag{1.20}$$

So, if the solution function is literally known, then the state transition matrix can also be written in closed form. Notice that this literally states that the state transition matrix is the derivative of the state at time t with respect to the state at time t_o. The two body problem is one system for which both the closed form state solution and the closed form state transition matrix are available.

For the common case where a solution is not available, formulating the variational equations involves finding $A(t)$. This involves N^2 partial derivatives of the equations of motion. For example, consider the two body problem in inertial rectangular coordinates. The state vector is $\mathbf{x}^T = (\mathbf{r}, \mathbf{v})$, and the equations of motion are

$$\dot{\mathbf{x}} = \begin{pmatrix} \dot{\mathbf{r}} \\ \dot{\mathbf{v}} \end{pmatrix} = \begin{pmatrix} \mathbf{v} \\ -\mu\mathbf{r}/r^3 \end{pmatrix} \tag{1.21}$$

Straightforward calculation of partial derivatives will show $A(t)$ to have the form

$$A(t) = \begin{pmatrix} \phi & I \\ A_{21} & \phi \end{pmatrix} \tag{1.22}$$

where ϕ is a 3 by 3 null matrix, and I is the identity matrix. The messy part of $A(t)$ involves the partials of the force with respect to the position vectors:

$$A_{21} = \begin{pmatrix} -\mu/r^3 + 3\mu x^2/r^5 & 3\mu xy/r^5 & 3\mu xz/r^5 \\ 3\mu yx/r^5 & -\mu/r^3 + 3\mu y^2/r^5 & 3\mu yz/r^5 \\ 3\mu zx/r^5 & 3\mu zy/r^5 & -\mu/r^3 + 3\mu z^2/r^5 \end{pmatrix} \tag{1.23}$$

For systems where the forces are more complex than the two body problem, this is the block of $A(t)$ which becomes correspondingly more and more messy. However, if the equations of motion can be written, then the equations of variation may be written also.

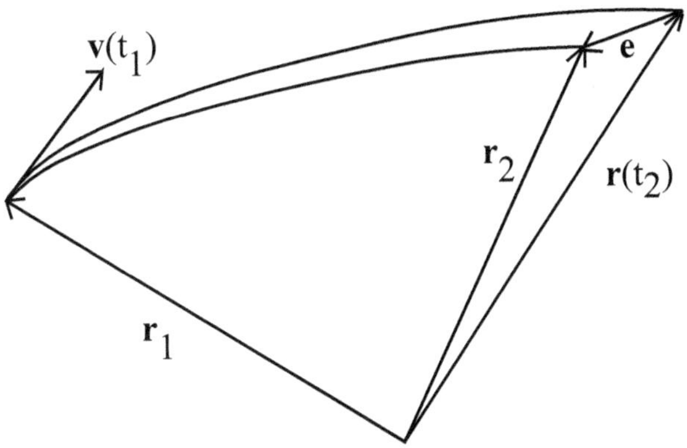

Figure 1.1: Two Position Vectors, Time of Flight

1.5 Boundary Value Problems

One very important class of problems often attacked numerically is the two point boundary problem. In the two body system the problem of orbit determination from two position vectors and the time of flight is one such problem. In well – studied systems like the two body problem, analytic or nearly analytic solutions may be available. The advent of real spaceflight makes it necessary to solve this type of problem in a fully realistic (and therefore complex) realization of the dynamics. In fact, this same problem will be discussed in this section, but for more general gravitational fields. The method described in this section is often called the "shooting method" by numerical analysts. While other methods have been developed and studied, one characteristic they all have in common is the need for a decent first guess.

Consider the two position vector / time of flight problem shown in Figure 1.1. We wish to leave position vector $\mathbf{r}_1$ at time t_1, and arrive at position vector $\mathbf{r}_2$ at time t_2. This is an extremely common problem that often arises in mission analysis and operational planning. The dynamics may be much more complex than the two body problem, in which case the two body problem solution may give (at best) a good first guess. In interplanetary trajectories, the method of patched conics serves a similar purpose. Now, take an approximate velocity at t_1, say $\mathbf{v}_1$, and use it with the known initial position vector $\mathbf{r}_1$ to propagate the trajectory to time t_2. At this point, we discover, of course, that the desired final position vector $\mathbf{r}_2$ is not exactly obtained, and there is an error vector

$$\mathbf{e} = \mathbf{r}_2 - \mathbf{r}(t_2) \tag{1.24}$$

where the second quantity is the position on the integrated trajectory. The problem is to fix the initial velocity vector $\mathbf{v}_1$ to obtain a zero error vector $\mathbf{e}$ at t_2.

Now, suppose that the state vector $\mathbf{x}$ was just the position and velocity vectors in rectangular coordinates. Then, the state transition matrix relationship

$$\delta\mathbf{x}(t_2) = \Phi(t_2, t_1)\delta\mathbf{x}(t_1) \tag{1.25}$$

can be written as

$$\begin{pmatrix} \delta\mathbf{r}(t_2) \\ \delta\mathbf{v}(t_2) \end{pmatrix} = \begin{pmatrix} \partial\mathbf{r}(t_2)/\partial\mathbf{r}(t_1) & \partial\mathbf{r}(t_2)/\partial\mathbf{v}(t_1) \\ \partial\mathbf{v}(t_2)/\partial\mathbf{r}(t_1) & \partial\mathbf{v}(t_2)/\partial\mathbf{v}(t_1) \end{pmatrix} \begin{pmatrix} \delta\mathbf{r}(t_1) \\ \delta\mathbf{v}(t_1) \end{pmatrix} \tag{1.26}$$

Now, in this expression, we can recognize that $\delta\mathbf{r}(t_2) \approx \mathbf{e}$, the error vector. Also, we are not allowed to change the initial position vector, so we must put $\delta\mathbf{r}(t_1) = 0$. The first three components of (1.26) then become

$$\mathbf{e} = \frac{\partial\mathbf{r}(t_2)}{\partial\mathbf{v}(t_1)}\delta\mathbf{v}(t_1) \tag{1.27}$$

where the partial derivative is the upper right 3 x 3 submatrix of the state transition matrix. This can be solved for the correction to the initial velocity vector

$$\delta\mathbf{v}(t_1) = \left(\frac{\partial\mathbf{r}(t_2)}{\partial\mathbf{v}(t_1)}\right)^{-1}\mathbf{e} \tag{1.28}$$

This correction is to be added to the initial velocity $\mathbf{v}_1$ to find a better initial velocity.

Now, in forming the relation for the state transition matrix, quadratic and higher order terms have been dropped. So, the next step is to repeat the integration, calculate a new error vector and state transition matrix, and continue until convergence is achieved. This would be performed as the initial targeting calculation for a lunar mission, for example. Once the spacecraft is on its way, however, and the trajectory is already *very close* to what we wish, then this method can converge in one iteration. This would be done for a mid-course trajectory correction, for example. Note that this type of algorithm is a multi–dimensional version of the Newton–Rhapson method.

The method can be extended to more complex sets of end conditions. Consider the lunar return trajectory shown in Figure 1.2. We still wish to determine the velocity vector $\mathbf{v}$ at t_1 to leave lunar orbit, but the final conditions are not as simple. Since we can change three initial conditions (the components of $\mathbf{v}_1$) we can specify three final conditions. These can be any combinations of the final position and velocity vectors. For Apollo spacecraft, the entry angle at

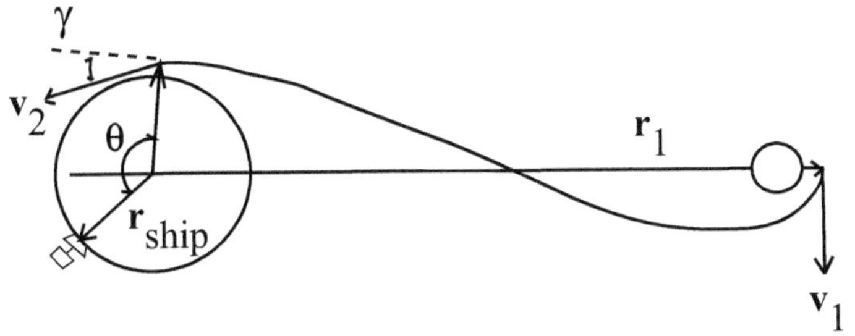

Figure 1.2: Apollo Return Trajectory

the beginning of reentry had to be $\gamma_o = -7.5^o$, with a very small tolerance. So, our first condition is

$$\sin \gamma_o = \frac{\mathbf{r}_2 \cdot \mathbf{v}_2}{|\mathbf{r}_2||\mathbf{v}_2|} \tag{1.29}$$

The reentry must begin at a given altitude, about $h_o = 300,000$ ft, so our second condition is

$$h_o = |\mathbf{r}_2| - R_\oplus \tag{1.30}$$

Finally, if the recovery squadron is located at $\mathbf{r}_{ships}$, it would be nice to bring the capsule down near them. If the total reentry covers an angle θ across the earth's surface (over 1/4 the earth's circumference for Apollo), then our third condition is

$$\cos \theta = \frac{\mathbf{r}_{ships} \cdot \mathbf{r}_2}{|\mathbf{r}_{ships}||\mathbf{r}_2|} \tag{1.31}$$

Actually, there are still insufficient quantities specified here, since we would like the recovery squadron to be in the orbital plane, also.....Apollo capsules had only limited crossrange capability. However, the squadron has almost three days from the departure from lunar orbit to get under this plane.

Now, group all of the end conditions into a three component vector $\mathbf{G}$:

$$\mathbf{G} = \begin{pmatrix} |\mathbf{r}_2| - R_\oplus - h_o \\ \mathbf{r}_2 \cdot \mathbf{v}_2 - |\mathbf{r}_2||\mathbf{v}_2| \sin \gamma_o \\ \mathbf{r}_{ships} \cdot \mathbf{r}_2 - |\mathbf{r}_{ships}||\mathbf{r}_2| \cos \theta \end{pmatrix} \tag{1.32}$$

On the desired trajectory, we would have $\mathbf{G} = \mathbf{0}$.

Now, we can start at $\mathbf{r}_1$ with a guessed $\mathbf{v}_1$ and propagate the trajectory and Φ matrix to time t_2. At this point, we can calculate the $\mathbf{G}$ vector, and find that it is not zero, but has a nonzero value $\mathbf{G} = \mathbf{e}$. Now, calculate the 3 x 6 matrix

$\partial\mathbf{G}/\partial\mathbf{x}$, and evaluate it at t_2. Small changes $\delta\mathbf{x}(t_2)$ produce small changes in $\mathbf{G}$, $\delta\mathbf{G}$, according to

$$\delta\mathbf{G} = \left.\frac{\partial\mathbf{G}}{\partial\mathbf{x}}\right|_{t_2} \delta\mathbf{x}(t_2) \tag{1.33}$$

Let us abbreviate the partials matrix $\partial\mathbf{G}/\partial\mathbf{x}$ as B. Combining (1.33) with the propagation law for $\delta\mathbf{x}$, we obtain

$$-\mathbf{e} \approx \delta\mathbf{G} = B\delta\mathbf{x}(t_2) = B\Phi(t_2, t_1)\delta\mathbf{x}(t_1) \tag{1.34}$$

The product $B\Phi$ is again a 3 x 6 matrix, the first 3 x 3 submatrix giving the change in $\mathbf{G}$ with changes in $\mathbf{r}_1$ (which we are not allowed to change), while the second 3 x 3 submatrix gives the change in $\mathbf{G}$ with changes in $\mathbf{v}_1$

$$B\Phi = \begin{pmatrix} \dfrac{\partial\mathbf{G}}{\partial\mathbf{r}(t_1)} & \dfrac{\partial\mathbf{G}}{\partial\mathbf{v}(t_1)} \end{pmatrix} \tag{1.35}$$

So, the correction to the approximate velocity vector $\mathbf{v}_1$ at transearth injection is given by

$$\delta\mathbf{v}_1 \approx -\left(\frac{\partial\mathbf{G}}{\partial\mathbf{v}(t_1)}\right)^{-1} \mathbf{e} \tag{1.36}$$

since we wish to introduce a change at time t_2 that will cancel the current error $\mathbf{e}$. Again, if this is the initial maneuver calculation, we would need to iterate the entire process to obtain convergence. If time t_1 is some point along the return trajectory, this same algorithm would furnish the midcourse correction burn with only a few iterations. In this case, if we start with an initial guess equal to our actual velocity, then the necessary maneuver is given by

$$\Delta\mathbf{v}(t_1) = \sum_j \delta\mathbf{v}_{1j} \tag{1.37}$$

after the iterations j have converged.

Consider the problem of constructing the Apollo "free - return" trajectory shown in Figure 1.3. This type of trajectory would have returned a disabled Apollo spacecraft to earth reentry after a close lunar flyby. Here the initial position is specified (it is where we are), and the initial velocity must be altered to effect a safe return to the earth and reentry conditions. Apollo spacecraft left low altitude earth parking orbit already riding the free return trajectory. The nearly disastrous Apollo 13 mission had just maneuvered off the free return trajectory when the explosion of an oxygen tank disabled the spacecraft, forcing a return to (nearly) the original orbit. This is just the same boundary problem, but starting at a general point in the earth–moon system.

It is even possible to extend the shooting method to the case where combinations of the initial conditions can be specified as well. Boundary value problems, when properly posed, always have a number of specified final conditions equal to the number of unknown initial conditions. When first encountered, the posing of boundary value problems appears to be an art form, but your ability to do it improves with practice.

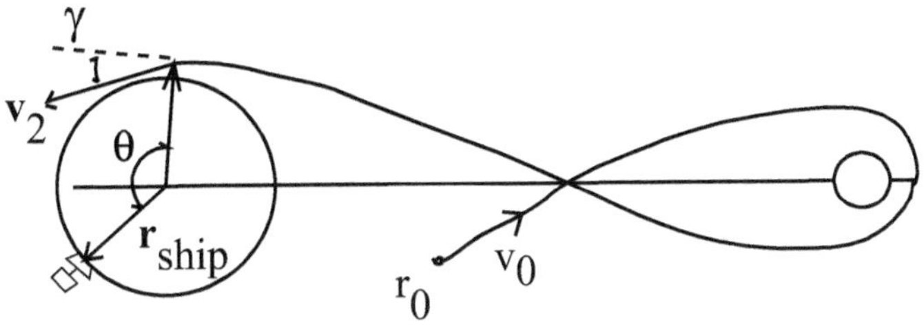

Figure 1.3: Apollo Free Return Trajectory

1.6 Periodic Orbits

Numerical integration can supply information on a specific orbit, for example, an Apollo lunar transfer trajectory. However, these trajectories typically are of interest for only a short interval of time. One difficulty with numerically integrated orbits is that they usually do not supply any information on the long term behavior of an orbit. Henri Poincaré has said that the primary problem in celestial mechanics is the behavior of orbits as time $t \to \infty$. Normally with numerical integration, as $t \to \infty$ the quantity $T \to \infty$, where T is your computer execution time. Also, as the time of integration grows, roundoff error tends to grow (typically as the square root of the number of integration steps), and the results become progressively less trustworthy.

However, there is one case where numerical integration can supply information on the behavior of a system as $t \to \infty$. This is the case of a periodic orbit. If the orbit closes on itself after time τ, if the velocity vector also returns to the same value after time τ, and if the external forces in the system are also the same, then the system motion from time τ to 2τ is simply a repeat of the motion from 0 to τ. The motion from 2τ to 3τ is a second repeat of the motion in the initial interval. So, for a finite numerical integration, information is gained on the behavior as $t \to \infty$. Typically, periodic orbits only exist in systems which do not contain the time in their Hamiltonians (for example, the restricted three body problem), or which are themselves periodic in time.

The problem of finding a periodic orbit is simply a two point boundary value problem, where the two points happen to be the same point. Integrate the orbit $\mathbf{x}(t)$ and the state transition matrix Φ from 0 to τ. Now, with a guessed set of initial conditions, it is unlikely that $\mathbf{x}(0) = \mathbf{x}(\tau)$. There will be an error vector

$$\mathbf{e} = \mathbf{x}(\tau) - \mathbf{x}(0) \tag{1.38}$$

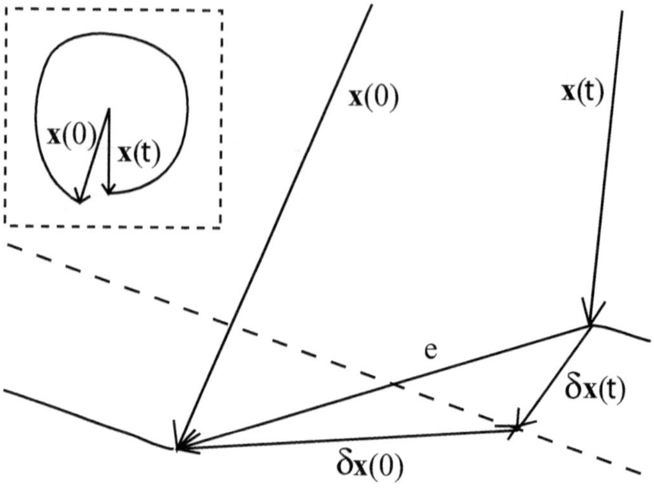

Figure 1.4: Periodic Orbit Boundary Conditions

as shown in Figure 1.4. Now, we wish to determine corrections to the initial conditions $\mathbf{x}(0)$ to close the orbit at $t = \tau$. A change in the initial conditions $\delta\mathbf{x}(0)$ will produce a change

$$\delta\mathbf{x}(\tau) = \Phi(\tau, 0)\delta\mathbf{x}(0) \tag{1.39}$$

at the end of one period. To get the orbit to close, we need

$$\mathbf{e} = \delta\mathbf{x}(0) - \delta\mathbf{x}(\tau) \tag{1.40}$$

If we eliminate $\delta\mathbf{x}(\tau)$ in (1.39), we obtain

$$\delta\mathbf{x}(\tau) = -\mathbf{e} + \delta\mathbf{x}(0) = \Phi(\tau, 0)\delta\mathbf{x}(0) \tag{1.41}$$

This can be rearranged to yield

$$\delta\mathbf{x}(0) = -\left(\Phi(\tau, 0) - I\right)^{-1}\mathbf{e} \tag{1.42}$$

Which gives corrections to the initial conditions necessary to ensure that the orbit will close.

However, the computation of periodic orbits is not always this simple. In the case of systems which are independent of time, the Hamiltonian is independent of time, and is therefore a constant of the motion. In this case, one eigenvalue of the state transition matrix $\Phi(\tau, 0)$ will always be 1. This means that the inverse matrix required in (1.42) is singular. There are various approaches which can be taken at this point, usually based on the symmetries of the problem. This problem occurs when constructing periodic orbits in the restricted problem of three bodies. The equations of motion are

$$\begin{aligned} \dot{x} &= v_x, \quad \dot{y} = v_y \\ \dot{v}_x &= 2v_y + x - \frac{(1-\mu)(x-\mu)}{r_1^3} - \frac{\mu(x+1-\mu)}{r_2^3} \\ \dot{v}_y &= -2v_x + y - \frac{(1-\mu)y}{r_1^3} - \frac{\mu y}{r_2^3} \end{aligned} \tag{1.43}$$

where the radii are

$$r_1 = \sqrt{(x-\mu)^2 + y^2}, \quad r_2 = \sqrt{(x+1-\mu)^2 + y^2} \tag{1.44}$$

The $A(t)$ matrix calculation is simple in principle but somewhat painful in practice.

Notice from the equations of motion that the problem is symmetric for reflection in the x axis. That is, if we write $x \Rightarrow x', y \Rightarrow -y', v_x \Rightarrow -v_x', v_y \Rightarrow v_y'$, then the equations of motion are unchanged if time runs backwards ($t \Rightarrow -t'$). For every trajectory there is a mirror image trajectory in the x axis which is running backwards in time. We can combine two such trajectories at any point where they both cross the x axis perpendicularly.

So, to find a symmetric periodic orbit in the restricted problem of three bodies, we need only integrate for one half of the orbital period. Furthermore, if we start at one of the perpendicular x axis crossings, we already know that $v_x = 0$ and $y = 0$, so only x_o and v_{yo} are unknown initial conditions. At the next crossing of the x axis, we can ensure that the orbit will be symmetric (and therefore periodic) if when $t = \tau/2$, we have $y = 0$ and $v_x = 0$ also. So, it becomes a reasonably simple matter to extract the four elements of Φ to pose this boundary value problem.

1.7 Floquet Theory

Once a periodic orbit has been calculated, you have a special solution to the full system equations of motion. Linearizing about the periodic orbit, the variational equations

$$\delta\dot{\mathbf{x}} = A(t)\delta\mathbf{x} \tag{1.45}$$

become time periodic linear differential equations. The case of time periodic linear systems is only slightly more difficult than the constant coefficient linear systems which arise when we linearize about an equilibrium point. One of the most important facts to know about an equilibrium point is its stability information, which is obtained by solving the linear variational equations near the equilibrium point. Similarly, linearizing about a periodic orbit produces periodic coefficient linear systems, and as discovered by Floquet, these can also be solved, giving stability information for the periodic orbit.

The main result of Floquet is that the state transition matrix (the solution matrix for equation (1.45)) can be written in the form

$$\Phi(t,0) = F(t)e^{Jt}F^{-1}(0) \tag{1.46}$$

where $F(t)$ is a periodic matrix, and J is a Jordan normal form matrix of the system frequencies. Usually J is diagonal, and its' diagonal elements are termed Poincaré exponents. Equation (1.46) for time periodic systems can be compared to the usual form of the state transition matrix for constant coefficient systems:

$$\Phi(t,0) = Fe^{Jt}F^{-1} \tag{1.47}$$

where J is again the (usually) diagonal matrix of system eigenvalues, and F is the (constant) matrix of the eigenvectors of A. So, the only difference in the case of periodic coefficient systems is that the eigenvector matrix F becomes a periodic function of time.

However, the method of solving for J and F differs for a periodic system. Begin with the matrix $\Phi(\tau,0)$, evaluated at one period. Usually the only method for obtaining $\Phi(\tau,0)$, termed the monodromy matrix, is to numerically integrate (1.45) for one period. This is usually done as a by–product of the calculation of the periodic orbit itself. Evaluating (1.46) at the end of one period, we have

$$\Phi(\tau,0) = F(0)e^{J\tau}F^{-1}(0) \tag{1.48}$$

remembering that $F(\tau) = F(0)$, since F is periodic. This can be rearranged to yield

$$F^{-1}(0)\Phi(\tau,0)F(0) = e^{J\tau} \tag{1.49}$$

This explicitly states that $F(0)$ is the eigenvector matrix of $\Phi(\tau,0)$, so we obtain the eigenvector matrix F at $t = 0$ by calculating the usual eigenvectors of the monodromy matrix. Also, if λ_i are the eigenvalues of $\Phi(\tau,0)$, (also called the characteristic multipliers), then (1.49) states that

$$\lambda_i = e^{\omega_i \tau} \tag{1.50}$$

where the ω_i, the diagonal elements of J (or the Poincaré exponents) are

$$\omega_i = \frac{1}{\tau}\ln\lambda_i \tag{1.51}$$

The Poincaré exponents can be interpreted just like the eigenvalues of a constant coefficient system. The imaginary part of ω_i is the oscillatory frequency of the mode i, while a positive real part indicates instability. Of course, the usual theorem about the ω_i appearing as either real numbers or complex conjugate pairs applies. However, for Hamiltonian systems, there is one additional extension: the ω_i must also appear as positive / negative pairs. This means that, for Hamiltonian systems without dissipation, the only possible cases are a pair of positive / negative real numbers, or a pair of positive / negative pure imaginary numbers. At this point, then, we have complete linear stability information for the periodic orbit. It is quite common to stop at this point in a Floquet solution.

However, we do not yet have the complete solution, since we only have the periodic matrix $F(t)$ at $t = 0$. To complete the solution we must obtain this periodic matrix function over one period. If we substitute the solution form (1.46) into the differential equation it satisfies, equation (1.45), we obtain

$$\dot{F} = A(t)F(t) - F(t)J \tag{1.52}$$

This is a matrix set of differential equations for $F(t)$, for which we found initial conditions in (1.49). So, one last integration from $t = 0$ to $t = \tau$ will furnish the eigenvector matrix $F(t)$ over one period. Since (1.52) is very similar in form to (1.45), usually the same code can be adapted to perform both functions. Periodic functions of time can be efficiently reduced to their Fourier series coefficients by the method of harmonic analysis, covered in Brouwer and Clemence.

Periodic orbits thus furnish a method of numerically attacking an unsolvable dynamical problem with a technique that is one level more complex than searching for equilibrium points. In both cases, the solution to a set of linear equations furnishes stability information. Periodic linear systems also occur in satellite stability and helicopter dynamics.

1.8 Lyapunov Exponents

In the general case of the variational equations, it is still possible to ask stability questions, although a general solution algorithm is not available. Consider numerically integrating a solution to

$$\delta\dot{\mathbf{x}} = A(t)\delta\mathbf{x} \tag{1.53}$$

with some random initial condition $\delta\mathbf{x}(t_o)$. If this solution is growing exponentially at some average rate λ_1, comparing norms of the solution at two times gives

$$|\delta\mathbf{x}(t)| = |\delta\mathbf{x}(t_o)| \exp(\lambda_1(t - t_o)) \tag{1.54}$$

Rearranging gives

$$\lambda_1(t - t_o) = \frac{1}{t - t_o} \ln \frac{|\delta\mathbf{x}(t)|}{|\delta\mathbf{x}(t_o)|} \tag{1.55}$$

The quantity λ_1 is a function of the time interval, and only an average over a long time has meaning. As we let $t - t_o$ grow, two things happen. First, we obtain a more accurate value for the average rate of exponential growth. We would really like to calculate the limit

$$\lambda_1(\infty) = \lim_{t\to\infty} \frac{1}{t - t_o} \ln \frac{|\delta\mathbf{x}(t)|}{|\delta\mathbf{x}(t_o)|} \tag{1.56}$$

But the second thing that happens as $t - t_o$ grows is that $\delta\mathbf{x}(t)$ will overflow within the computer.

To prevent the overflow, we can use the basic linearity property of (1.53) to renormalize the solution. If the magnitude of $\delta\mathbf{x}$ is inconveniently large, we are free to renormalize it to a unit vector. During the interval in which $\delta\mathbf{x}$ was growing, its average exponential rate of growth was given by (1.55). Over many such intervals $(t_i,\ t_{i+1})$, the overall rate of exponential increase after n time intervals will be

$$\lambda_1 = \frac{1}{t_n - t_o}\sum_{i=1}^{n} \ln \frac{|\delta\mathbf{x}(t_i)|}{|\delta\mathbf{x}(t_{i-1})|} \tag{1.57}$$

Normalizing after each interval keeps the solution from overflowing, while the overall logarithmic rate of increase is still calculable.

But this only calculates *one* Lyapunov exponent: the largest one. There should be N Lyapunov exponents in an Nth order system. The most obvious thing to try in order to calculate the others is to try other initial conditions for $\delta\mathbf{x}(t_o)$. Unfortunately, the direction in space with the largest Lyapunov exponent is attractive: all solutions will eventually converge to this one solution. In order to explore the behavior of the other directions in space, it is necessary to be more careful.

The area spanned by two solutions $\delta\mathbf{x}_1$ and $\delta\mathbf{x}_2$ should grow exponentially at the rate $\lambda_1 + \lambda_2$, where λ_2 is the *second* largest Lyapunov exponent. Of course, the second solution $\delta\mathbf{x}_2$ will converge towards the first. This is just another version of the overflow problem with the first solution. Periodically we can normalize the first solution, as earlier, and then take the second solution and i) remove any component it has along the first solution, and then ii) normalize the result. So at intervals of time, we

$$\delta\mathbf{x}_{1,new} = \delta\mathbf{x}_1 / |\delta\mathbf{x}_1| \tag{1.58}$$

$$\delta\mathbf{x}_{2,new} = \frac{\{\delta\mathbf{x}_2 - \delta\mathbf{x}_2 \cdot \delta\mathbf{x}_{1,new}\}}{|\delta\mathbf{x}_2 - \delta\mathbf{x}_2 \cdot \delta\mathbf{x}_{1,new}|} \tag{1.59}$$

In a third order system, we subtract any contribution of $\delta\mathbf{x}_3$ along either $\delta\mathbf{x}_{1,new}$ and $\delta\mathbf{x}_{2,new}$, and then normalize the result

$$\delta\mathbf{x}_{3,new} = \frac{\left\{\delta\mathbf{x}_3 - \sum_{j=1}^{2} \delta\mathbf{x}_3 \cdot \delta\mathbf{x}_{j,new}\right\}}{\left|\delta\mathbf{x}_3 - \sum_{j=1}^{2} \delta\mathbf{x}_3 \cdot \delta\mathbf{x}_{j,new}\right|} \tag{1.60}$$

This is just the Gram–Schmidt orthonormalization process for a set of vectors. It "preserves the subspaces" spanned by any set of the solutions before and after orthonormalization, so when we continue the integration, we will still be sampling the same length, area, volume, etc as before.

The one, two, and three dimensional areas are just the denominators in (1.58), (1.59), and (1.60). So, we keep track of the sums

$$\lambda_1 = \frac{1}{t_n - t_o}\sum_{i=1}^{n} \ln D_1$$

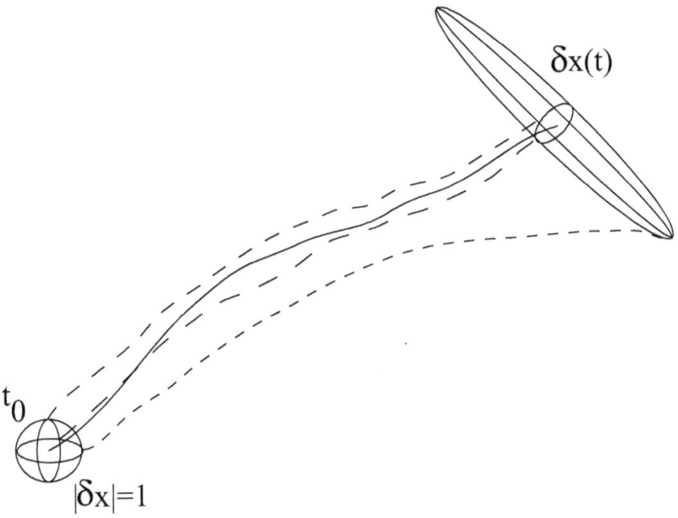

Figure 1.5: Extremal trajectories in a Lyapunov exponent calculation

$$\lambda_1 + \lambda_2 = \frac{1}{t_n - t_o} \sum_{i=1}^{n} \ln D_2 \tag{1.61}$$

$$\lambda_1 + \lambda_2 + \lambda_3 = \frac{1}{t_n - t_o} \sum_{i=1}^{n} \ln D_3$$

where the D_k are the divisors in the Gram–Schmidt process. As $t_n \to \infty$, these expressions converge to the Lyapunov exponents.

There is a simple geometric interpretation of the development above, and that to follow. We are dealing with a linear system, and dividing by the magnitude of the initial conditions, so we may as well consider all initial conditions $\delta\mathbf{x}$ which have unit magnitude. A linear system is scale invariant. Also, we are using the Gram–Schmidt process to seek orthonormal sets of initial conditions. The particular set of initial conditions we are seeking is the set that grows (or shrinks) by the largest amount(s) at time t. This is shown schematically in Figure 1.5, where the sphere $|\delta\mathbf{x}(t_0)| = 1$ of all possible initial displacements is deformed into the ellipsoid of final $\delta\mathbf{x}(t)$. The special trajectories represent the principal axes of the ellipsoid.

The above algorithm is stable in the case where we allow $t \to \infty$. For shorter intervals, there is another equivalent method, which unfortunately is unstable for long time intervals. The Lyapunov exponents are the extrema of

$$\mathcal{J} = \frac{1}{t - t_0} \ln \left\{ \frac{|\delta\mathbf{x}(t)|}{|\delta\mathbf{x}(t_0)|} \right\} \tag{1.62}$$

over the initial conditions $\delta\mathbf{x}(t_0)$. Since (1.53) is a linear system, we can restrict ourselves to initial conditions with unit norm. Also, it is easier to deal with the norm *squared*, rather that with the square root. This leads to the slightly altered constrained optimization problem

$$\begin{aligned} \mathcal{J}' &= \delta\mathbf{x}(t)^T\delta\mathbf{x}(t) - \mu\left(\delta\mathbf{x}(t_0)^T\delta\mathbf{x}(t_0) - 1\right) \\ &= \delta\mathbf{x}(t_0)^T\Phi^T\Phi\delta\mathbf{x}(t_0) - \mu\left(\delta\mathbf{x}(t_0)^T\delta\mathbf{x}(t_0) - 1\right) \end{aligned} \tag{1.63}$$

where μ is a Lagrange multiplier. The Lagrange multiplier allows us to ignore the question of which component of $\delta\mathbf{x}$ is the one that is being constrained. Then, notice that the gradient of a dot product is given by

$$\frac{\partial}{\partial x}\mathbf{a}^T\mathbf{x} = \frac{\partial}{\partial x}\mathbf{x}^T\mathbf{a} = \mathbf{a} \tag{1.64}$$

and the quadratic form (1.63) is essentially two such dot products. Then, calculating the partial derivatives of the quadratic form (1.63) component by component leads to the linear problem

$$\Phi^T\Phi\delta\mathbf{x} + (\delta\mathbf{x}^T\Phi^T\Phi)^T - 2\mu\delta\mathbf{x} = 0 \tag{1.65}$$

and using the matrix identity $(AB)^T = B^TA^T$ yields

$$\left\{\Phi^T\Phi - \mu I\right\}\delta\mathbf{x}(t_0) = 0 \tag{1.66}$$

This is an eigenvalue problem on the matrix $\Phi^T\Phi$, or the *singular value decomposition* of Φ. Comparing the form of (1.66) to the expected behavior $|\delta x(t)| = |\delta x(t_0)|\exp(\lambda(t-t_0))$ allows us to show that $\mu = \lambda^2(t-t_0)^2$. Standard software can be used to calculate the Lyapunov exponents over short time intervals, before the differences in the different growth rates $\lambda_i(t-t_0)$ exceeds the word length of the computer. Then progressively, the smallest Lyapunov exponents become uncalculable, until only the largest remains.

Now, we are familiar with the solution to the constant coefficient case, where the eigenvalues of the A matrix, ω_i are complex. Their real parts govern stability, while their imaginary parts are oscillation frequencies. In the periodic coefficient case, the Poincaré exponents ω_i are again complex, with the real part governing stability and the imaginary parts being oscillation frequencies. In the general case, Lyapunov exponents λ_i *are purely real.* There is, as yet, no definition which permits an imaginary part.

1.9 Order and Chaos

You have probably noticed by now the reverence dynamicists have for integrals of the motion. This is not just a coincidence, since if an Nth order dynamical system possesses N independent integrals of the motion

$$\mathcal{I}_i = \mathcal{I}_i(q_j, p_j, t), \quad i = 1, N \quad j = 1, N/2 \tag{1.67}$$

then, in principle, these N equations can be solved for the N unknowns q_j, p_j to give

$$q_j(t) = q_j(\mathcal{I}_i, t), \quad p_j(t) = p_j(\mathcal{I}_i, t) \tag{1.68}$$

These equations give the system coordinates and momenta in terms of the time and N arbitrary constants $\mathcal{I}_i$. In other words, *equations (1.68) constitute the full solution to the original dynamical problem.*

The existence of a full set of integrals of the motion is thus equivalent to having the full solution to the problem. However, there seems to be a great gulf between problems like the harmonic oscillator and the two body problem, where integrals of the motion are everywhere; and problems like the general N body problem where the number of integrals is limited, or there may seem to be none at all. The N body gravitational problem possesses 10 integrals of the motion. These are center of mass linear momentum (6), total angular momentum (3), and total energy (1). A total of $6N$ integrals would be needed to solve the problem, so this is not even enough integrals to cover the case $N = 2$, which we know has more than 10 integrals.

Henri Poincaré was particularly interested in the restricted problem of three bodies, for several reasons. In a rotating frame, and using standard non – dimensional units, it is a Hamiltonian system:

$$H = \frac{1}{2}\left(p_x^2 + p_y^2\right) + yp_x - xp_y - \frac{1-\mu}{r_1} - \frac{\mu}{r_2} \tag{1.69}$$

where

$$r_1 = \sqrt{(x-\mu)^2 + y^2}, \quad r_2 = \sqrt{(x+1-\mu)^2 + y^2} \tag{1.70}$$

The planar restricted three body problem would require 4 integrals of the motion to allow its complete solution, but Poincaré was able to prove that the only analytic integral this system has is the Hamiltonian itself. Any other integrals, if they exist, are *not* analytic functions of the system coordinates, momenta, and the time.

This did not stop Poincaré from wondering about the existence of other integrals. He invented an analytic technique, the *surface of section,* which has since found wide application in numerical exploration. His reasoning goes as follows: the restricted problem has a phase space of dimension 4, and one known integral of the motion. Suppose we just consider all orbits with a specified value of the Hamiltonian, $H = H_o$. This group of orbits must fill a three dimensional subspace of the full four dimensional phase space. Now, if there were a second integral of the motion, it would further restrict the motion of the orbits with $H = H_o$. Each orbit on this manifold would itself be constrained to lie on a *two dimensional* surface imbedded in the three dimensional space $H = H_o$. On the other hand, if there is no second integral of the motion, then all orbits with $H = H_o$ can intertwine freely in the three dimensional subspace, and will eventually fill it up completely. There is a simple way to examine the internal structure of this three dimensional subspace: you simply slice it.

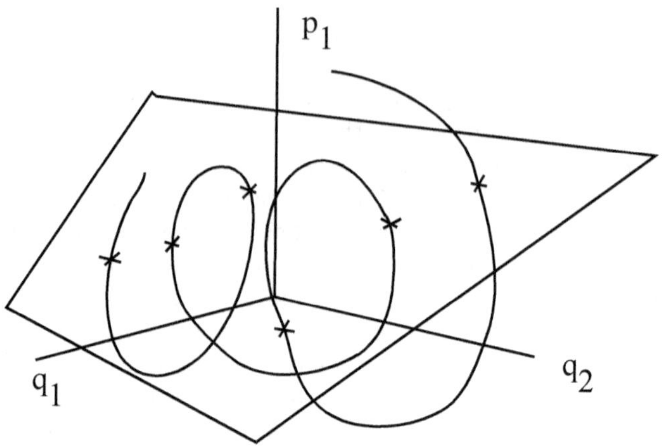

Figure 1.6: The Surface of Section

Imagine running a plane through this three dimensional space in some quite arbitrary way, as sketched in Figure 1.6. This plane is called a surface of section. Now, numerically integrate a trajectory with $H = H_o$, and every time the trajectory crosses the plane, you plot a point. If no second integral of the motion exists, then the plotted points will seem to fill an area on the surface of section, since the trajectory fills a three dimensional volume in phase space. On the other hand, if there is a second integral of the motion, this trajectory is constrained to lie on the two dimensional surface implicitly described by

$$H = H_o(q_i, p_i), \quad \mathcal{I}_{2o} = \mathcal{I}_2(q_i, p_i) \tag{1.71}$$

for some value $\mathcal{I}_{2o}$ of the second integral. The surface of section will slice this two dimensional surface in a series of lines, and the intersection points will appear to outline a series of closed curves on the plane of section. There are no other possibilities, so which is correct?

Real advances in science often occur when you realize that the question you are asking is nonsense. Students of quantum mechanics often are perplexed as to whether the electron is a particle or a wave. This question is equivalent to asking if a peach is an apple or an orange.....its neither, its a peach! In this case, the surprise waiting for Poincaré was that *both behaviors described above occur at the same time!*

Henri Poincaré was able to describe in 1892 the actual structure of the phase space of a non – integrable system. The widespread appreciation of his discovery was delayed because Poincaré said that the structure was too complicated to draw, and so he didn't try. It required almost 90 years for computer technology to advance to the point where numerical fools could rush in where theoretical angels feared to tread.

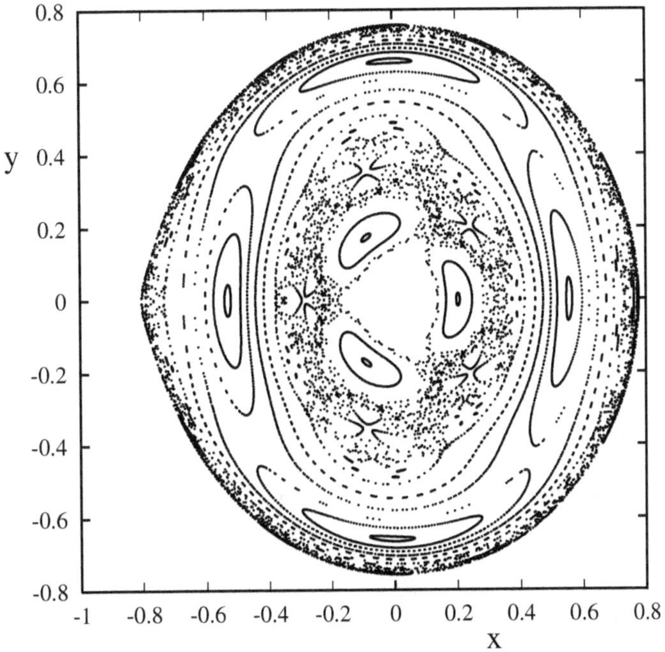

Figure 1.7: A Numerically Computed Surface of Section

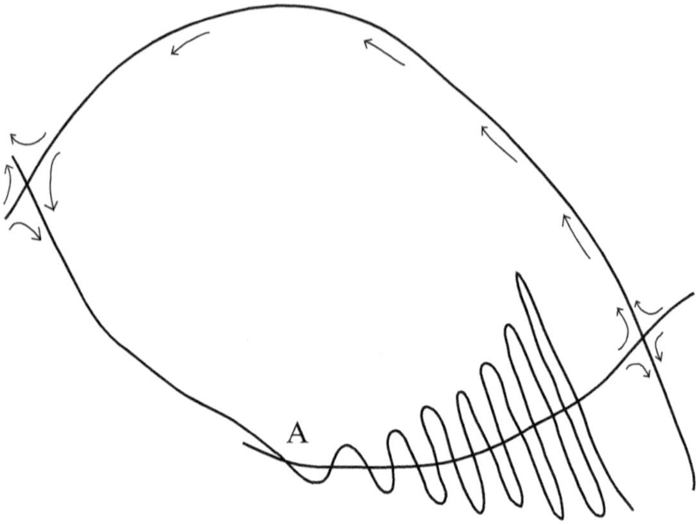

Figure 1.8: Wild Oscillations when Separatrices Intersect

Figure 1.7 shows a numerically calculated surface of section. In a sense, the problem it comes from is not important, since all "unsolvable" systems seem to exhibit many of the same behaviors. Large areas of the surface are covered by orderly, quite regular appearing curves. Usually, the system will have values of the Hamiltonian where the entire surface appears to be well behaved. Then, usually as the energy is increased, the alternate behavior begins to appear. This appears as "scruffy" areas where plotted points do not appear to form any nice pattern. In these areas, then, a second integral would seem to not exist, while in the "nice" areas its existence seems assured.

However, one further surprise awaits, as predicted by Poincaré. The breakdown occurs in the vicinity of "chain of island" structures. The center of each of these islands contains a point which cuts the surface of section repeatedly in just a finite number of places. By construction then, these trajectories must be periodic orbits, and stable periodic orbits, besides. In between the stable periodic "islands" are unstable "saddle point" periodic orbits. The surprise is that if the scale of the plot is increased, the coastline of the islands, which doesn't appear too clean, itself consists of more island chain structures, and *this hierarchical structure continues forever, at ever finer scales.*

The breakdown from order to chaos begins in the vicinity of unstable periodic orbits. These appear as saddle points on the surface of section, as sketched in Figure 1.8. The figure is a composite of two possible behaviors: both cannot occur at the same time. Along the top, the outgoing direction from the unstable periodic point joins smoothly with the incoming direction for another unstable periodic orbit. In regions where this behavior is the norm, phase space is split up into "island chain" structures, with alternating stable and unstable periodic orbits. It is easy to find areas on Figure 1.7 where this behavior seems dominant.

The lower half of Figure 1.8 shows the other possible behavior. The outgoing critical manifold from the unstable periodic orbit on the left crosses the incoming manifold for the next at a small angle, at point A. One crossing is enough to ensure that there will now be *an infinite number of crossings of the stable and unstable manifolds.* Because point A is on the incoming manifold of the periodic orbit to the right, it will approach that orbit, in the end, exponentially. So the one single intersection is mapped forward into an infinite sequence of intersections, getting closer and closer together as the unstable periodic orbit on the right is approached. Furthermore, for a Hamiltonian system, each loop will have the same area. As the bases of the loops become narrower, their amplitude becomes larger. Eventually, they become large enough that they extend all the way to the unstable orbit on the right.... and even further.

Our limited computational ability makes it impossible to follow this behavior for more than a very few cycles. The argument of the previous paragraph is easily extended to a proof, but it is an analytic argument. What appears on the surface of section is a broad area of "random" points: the single trajectory appears to fill an area, rather than forming a curve. Actually, of course, it is a curve, just an infinitely convoluted one. So "chaos" is really not random, it is simply infinitely complex order! Such an area is also easily found on Figure 1.7, near the chain of three islands at the center.

There is another way that the surface of section of a Hamiltonian system becomes infinitely complicated. At the core of an "island" structure is a stable periodic orbit, with pairs of imaginary Poincaré exponents. The frequency that the orbit "winds" around the central periodic orbit generally varies with amplitude, however. Near areas where the winding frequency is a rational multiple of the frequency of the orbit, a *resonance* is possible. Under very general conditions, there are special sets of initial conditions where a combination of the periodic orbit motion and the winding motion produce yet another periodic orbit. That is, an island structure can contain yet smaller island structures. They come into existence at the periodic orbit itself, initially with zero amplitude, in a process known as *bifurcation.* Again, just as with the intersection of critical curves shown in Figure 1.8, this process does not necessarily terminate. Each set of stable islands may contain yet smaller island / saddle chains, and so on, without limit. This is not necessarily chaotic motion per se, but it is infinitely complicated motion. It can be as dense as the rational numbers (numbers which are the ratio of two integers) are dense on the real numbers.

Further, there is a substantial difference in the behavior of trajectories in the "well behaved" and "chaotic" regions. Liouville's theorem for a Hamiltonian system states that the determinant of the state transition matrix Φ is always one. However, according to Lyapunov, the state transition matrix can be factored as

$$\Phi(t,0) = F(t)e^{Jt}F^{-1}(0) \tag{1.72}$$

even in the general time dependent case. The diagonal elements of J are in this case termed Lyapunov characteristic exponents. They are very difficult to compute. In well – behaved regions, however, the distance between two close trajectories grows no faster than linearly with time. This implies that the largest Lyapunov exponent is close to zero. On the other hand, in chaotic regions the distance between two close trajectories grows exponentially with time, and at least one Lyapunov exponent is positive and real.

Chaotic behavior makes prediction into the future or back into the past eventually pointless. If we have made an error $\mathbf{e}(t_o)$ in the initial value of the state vector, then the error we make predicting to some other time t is

$$\mathbf{e}(t) = \Phi(t,t_o)\mathbf{e}(t_o) \tag{1.73}$$

If the largest eigenvalue of the state transition matrix is growing exponentially with time, eventually our "small" initial error will grow "large". The asteroid (dead comet?) Charon orbits between Jupiter and Saturn at the present time, but it is in a chaotic trajectory. Every few thousand years it makes a close approach to one or the other object, and then its orbit alters radically. It is impossible to calculate where it came from, although capture of an incoming comet seems most probable. It is impossible to calculate where it will go, although ejection from the solar system is more probable than impact on a planet. By "impossible to calculate" I mean just that: knowing the current orbit to *hundreds of significant figures of precision* would barely suffice to propagate the orbit through even five close approaches. Perfectly deterministic equations of motion still can describe systems which, for any practical purpose, might as well be thrown dice. [2]

It would seem desirable for human craft to avoid such chaotic trajectories like the plague. However, all interplanetary orbits within our solar system lie within chaotic regions of the phase space. This makes them *very* sensitive to errors in navigation. However, there is another side to this problem. Rewrite equation (1.73) as

$$\delta\mathbf{x}(t_o) = \Phi^{-1}(t,t_o)\delta\mathbf{x}(t) \tag{1.74}$$

Now, this equation states that a large change at time t can be produced by a *very small* change in the state at time t_o. The incredibly rapid growth of $\Phi(t,t_o)$ can be used to produce drastic changes (usually at a planetary flyby) if the navigation is done to exquisite accuracy and if maneuvers are made early enough. The spectacularly successful mission of Voyager II (Jupiter – Saturn – Uranus – Neptune) is only the most obvious of the missions which are only possible because the trajectory is chaotic. The Galileo mission to Jupiter (by way of Venus, Earth, Earth, and two asteroids) also very successfully played this game of celestial billiards[3] with Jupiter's four large moons.

[2] Which are also allegedly a deterministic system!

[3] Billiards is also a deterministic system with chaotic trajectories.

The existence of chaos goes a long way towards salving our wounded pride at not being able to solve "simple" systems like the restricted problem of three bodies. If the structure of the phase space is infinitely complex and convoluted, it will defy description by simple functions like sine and cosine that we mere humans try to use. We are also beginning to discern an underlying structure and unity in chaos, however. Such apparently disparate phenomena as chaos in dynamics, turbulence in fluids, and the "randomness" of two cubical rigid bodies undergoing elastic collisions (dice) are apparently all the same. The richness of the game of the universe is that fairly simple rules lead to results of infinite complexity (eg, chess), instead of too simple rules leading to rapid boredom (eg, tic – tac – toe).

1.10 Manifold Dynamics

One of the most exciting developments in orbital mechanics has been the emergence of the study of *manifolds* in the solution space. The definition of a manifold is quite general and very esoteric, but we can think of a *differentiable manifold* as being a surface of lower dimension imbedded within the phase space. Usually, manifolds are studied for some characteristic that they share in common. Consider the short arcs of a stable and unstable periodic orbit shown in Figure 1.9. On the right, orbits near the stable periodic orbit undergo a sinusoidal oscillation about the periodic orbit as a guiding center. In the linear region, they lie on the surface of an elliptical tube, centered on the stable periodic orbit, and the axes of this elliptical tube can be found from the Floquet modal vectors associated with the pair of imaginary Poincaré exponents. Since the periodic orbit joins back on itself, this tube closes on itself, and looks like a donut in phase space. The mathematical term for this is a *torus*, and this tube is an example of what is termed a *center manifold* of the stable periodic orbit. If we sliced through a torus with a surface of section, we would observe an island structure.

Figure 1.9 also shows an unstable periodic orbit. In the linear regime, motion near the unstable periodic orbit is determined by the positive and negative pair of Poincaré exponents, and their associated modal vectors $\mathbf{f}$, the columns of the F matrix from section 1.7. There will be an incoming direction, determined by the modal vector $\mathbf{f}_i$ associated with the negative real Poincaré exponent, and an outgoing direction along the modal vector $\mathbf{f}_j$ associated with the positive real Poincaré exponent. A linear combination of the two of these makes the motion look like a saddle point, but moving along the unstable periodic orbit as a guiding center. Most nearby orbits depart from the vicinity of the unstable periodic orbit, but there are two special groups. The first group consists of all of the orbits that originate (at $t = -\infty$) on the unstable periodic orbit itself. These will take an "infinite" time to depart, but will do so along the direction of the outgoing modal vector $\mathbf{f}_j$. As this group of orbits move outward, they form a two dimensional manifold, the *unstable* manifold of the unstable periodic

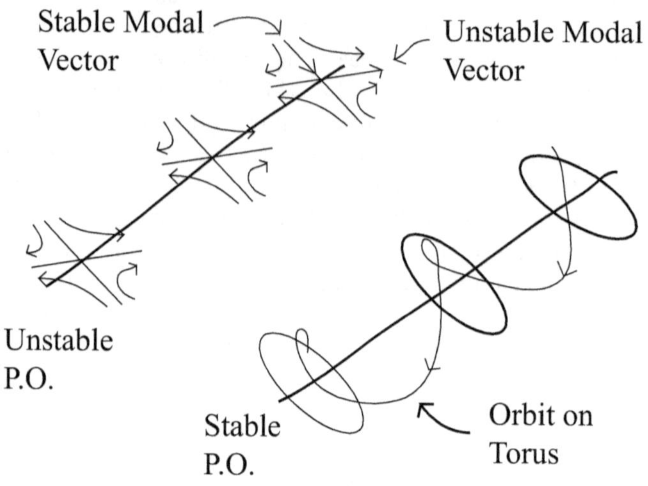

Figure 1.9: Motion Near Stable and Unstable Periodic Orbits

orbit. (The two dimensions are "distance outward" and phase along the periodic orbit.) Similarly, there is a select group of orbits with no component at all along the unstable mode, and these orbit wind up (as $t \to \infty$) on the periodic orbit itself. They can be found by propagating backwards in time, they form the *stable* manifold of the unstable periodic orbit. The stable manifold is also a two dimensional surface imbedded in the phase space. Since the restricted problem of three bodies is time–invariant, these are also termed *invariant manifolds.* They are always there, their position does not depend on the time of day.

As shown in Figure 1.10, these surfaces exist outside the linear region around the unstable periodic orbit. Sample orbits in either surface can be found by numerically integrating trajectories with initial conditions that consist of a point on the periodic orbit plus a small displacement along one of the Floquet modal vectors

$$\mathbf{X}(t_0) = \mathbf{X}_{PO}(t_0) + \epsilon \mathbf{f}_i \tag{1.75}$$

where ϵ is an appropriately small number. Of course, as shown in an earlier section, the unstable manifold of an unstable periodic orbit may connect smoothly into the stable manifold of another periodic orbit (see the upper half of Figure 1.8), in which case phase space will be divided into alternating stable and unstable periodic orbits in an island chain structure. Or, the stable and unstable manifolds of an unstable periodic orbit can intersect at a small angle, leading to wild oscillations. This is sketched in the lower half of Figure 1.8, and is visible near the center of Figure 1.7. But what happens in the really chaotic regions, when the stable and unstable manifolds of an unstable periodic orbit are not

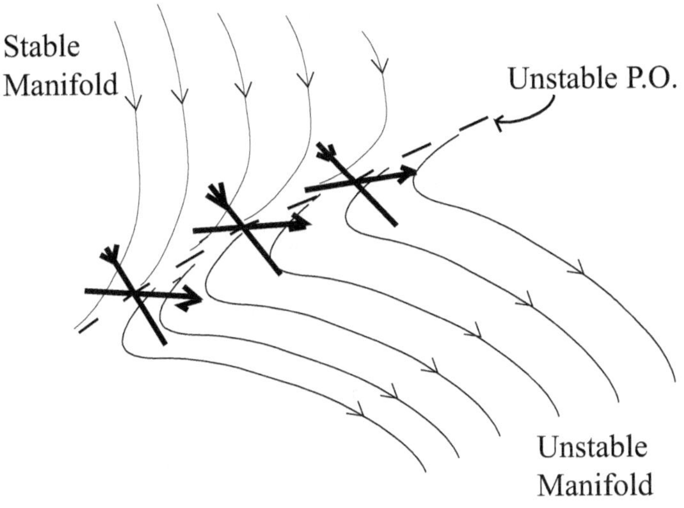

Figure 1.10: The Stable and Unstable Manifolds of an Unstable Periodic Orbit

"squeezed" between the tori (plural of torus) of stable periodic orbits?

First, numerically it is possible to explore this case, although theoretical predictions are not available. For the last thirty years, the United States has maintained satellites near the L_2 point of the Earth – Sun restricted three body problem. This places the spacecraft between the Earth and Sun, at about three times the distance to the Moon. From this position the spacecraft have served as early warning monitors for solar storms. Of course, a satellite placed directly at the L_2 point would appear at the center of the Sun's disc seen from the earth, and that would be very inconvenient for communications. The antenna on Earth would have to point *directly* at the sun. So, these satellites are placed into three dimensional "halo" orbits about the L_2 point. These are periodic orbits in the restricted problem, but are unstable. The satellite must maneuver occasionally to stay in their vicinity.

How does one maneuver back to an unstable periodic orbit? It could be done by using two maneuvers. The first would return the satellite's position to some point on the periodic orbit, and then the second maneuver would correct the satellite's velocity to put it back "on" the unstable periodic orbit. Inevitable small errors would give the satellite's state a small component along the unstable modal vector, making another maneuver necessary in the future, as the satellite attempts to leave, in an orbit essentially contained within the unstable manifold. Alternately, by changing the satellite's velocity it may be possible to maneuver directly from the satellite's current position within the unstable manifold to a corresponding position *within the stable manifold.* In a maneuver, of course, the satellites physical position does not change, only its velocity. But there may be places where a change in velocity makes a single maneuver jump from the unstable to the stable manifolds possible.

Now, early L_2 satellites were brought to their mission orbit by a long elliptical transfer to a point on the periodic orbit, and then a final maneuver inserted the spacecraft into the unstable periodic orbit. For the Genesis spacecraft, an entirely different approach was used. The *stable* manifold of the *unstable* periodic orbit is the set of all trajectories which will enter the actual periodic orbit as $t \to \infty$. Following this manifold backward in time, it was discovered that part of this manifold intersects the earth, and even at the longitude of the Cape Kennedy launch site. So, the Genesis spacecraft was directly inserted into the stable manifold of the unstable periodic orbit. No large second maneuver was necessary.

This requires precise navigation, of course. One could calculate the Lyapunov exponents of this arc of the stable manifold. It would not be surprising to find that trajectories in parts of the stable manifold tend to depart from that surface. In other words, the stable manifold of the unstable periodic orbit might itself be unstable. But accurate navigation and small maneuvers kept the spacecraft sufficiently close to this manifold that it entered the unstable periodic orbit.

The next part of the Genesis mission was truly spectacular. For the unstable manifold of the L_2 unstable periodic halo orbit intersects the stable manifold of another unstable periodic orbit at the Earth – Sun L_1 point. In other words, it is possible to transfer from the L_2 periodic orbit to the L_1 periodic orbit with no deterministic maneuver requirement. It can be done for free.... or as free as accurate navigation and maneuvering can make it. The unstable manifold of the L_1 periodic orbit again intersects the Earth, so the final part of the mission was to return Genesis' solar wind samples to Earth, from L_2, via L_1, all for "free". And this actually worked, although an unfortunate assembly error resulted in the "gee switch" being installed upside down, and the reentry capsule did not deploy the parachute at the last moment. The Genesis reentry capsule crashed, however, exactly where the orbital mechanics predicted.

The intersection of invariant manifolds is a new and exciting technique for mission planning in the solar system. Attention is focused on the moon systems of Jupiter, Saturn, and the solar system itself. Even when L_1 or L_2 manifolds do not intersect, they may come close enough to make sizable economies possible in interplanetary flight.

1.11 References

The stable algorithm for calculating the Lyapunov exponents is recent enough that the author feels compelled to cite A. Wolf, J.B. Swift, H.L. Swinney and J.A. Vastano in Physica D, volume 16, 1985.

We will cite the Brouwer and Clemence Fourier series method in full in a later chapter's appendix.

The exciting work being done in manifold dynamics already has a large number of contributors. The reader will continue to find the most current work in the conference proceedings of the Spaceflight Mechanics conferences hosted each year by the profession.

1.12 Problems

Problem 1. Find the $A(t)$ matrix for the restricted problem of three bodies, as given in Section 1.6, equations (1.43) and (1.44).

Problem 2. Prove that the restricted problem of three bodies possesses the reflection symmetry described in Section 1.6. Pose the algorithm for finding symmetric periodic orbits in this dynamical system. That is, which elements of $\Phi(\tau/2,0)$ are needed, and how are they used? Argue that you can construct the monodromy matrix for this orbit as

$$\Phi(\tau/2,-\tau/2) = \Phi(\tau/2,0)\Phi^{-1}(\tau/2,0)$$

Problem 3. The N body problem can be written in relative coordinates as (see 4.6)

$$\dot{\mathbf{X}} = \frac{d}{dt}\begin{pmatrix} \vec{\rho} \\ \dot{\vec{\rho}} \end{pmatrix} = \begin{cases} \dot{\vec{\rho}} \\ -\mu\vec{\rho}/\rho^3 - \sum_{j=3}^{N} \mu_j \left((\vec{\rho}-\vec{\rho}_j)/\rho_{2j}^3 + \vec{\rho}_j/\rho_j^3\right) \end{cases}$$

where $\rho_{2j} = |\vec{\rho}-\vec{\rho}_j|$, and the positions of the other planets $\vec{\rho}_j(t)$ are known functions of time. Find $A(t)$.

Problem 4. For the time periodic linear system, show that the inverse eigenvector matrix obeys the differential equation

$$\dot{F}^{-1} = -F^{-1}A(t) + JF^{-1}$$

Also, show that the time periodic transformation $\vec{\eta} = F^{-1}(t)\mathbf{x}$ reduces the time periodic system $\dot{\mathbf{x}} = A(t)\mathbf{x}$ to the constant coefficient, decoupled system $\dot{\vec{\eta}} = J\vec{\eta}$

Problem 5. Combinations of initial conditions can be handled in boundary value problems in much the same way that final conditions are implemented. Consider leaving the vicinity of the earth from a low earth orbit, as shown in Figure 1.11. You must maneuver tangentially from circular speed v_c by an amount δv at the correct point on the parking orbit ϕ to achieve given final conditions at lunar approach. Discuss how to handle the problem of finding the two unknown initial parameters $\mathbf{Z}^T = (\delta v, \phi)$ in a boundary value problem. *Hint:* Write the initial position and velocity vectors in terms of the parking orbit radius r_o, and $\delta v, \phi$.

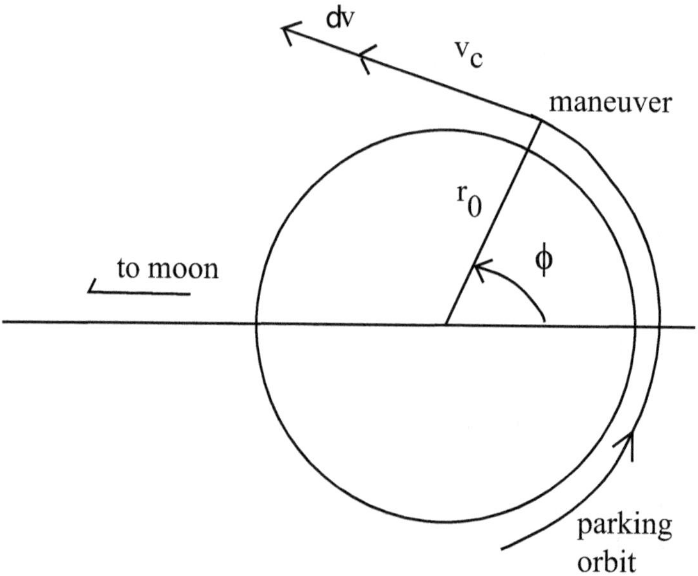

Figure 1.11: Tangential Departure from Low Earth Orbit

Chapter 2

Transformation Theory

2.1 Introduction

Most of modern perturbation theory in celestial mechanics is done in canonical variables using canonical transformation theory. Transformations of the system Hamiltonian to new sets of variables with "better" properties can not only simplify the equations of motion, but they can also *solve* the equations of motion. Even when a transformation to a set of variables solving the problem cannot be found, a new set of variables can often be found which solves the dominant part of the system. In celestial mechanics, the part of the system which can be solved is usually the two body motion portion, and the remainder of the system is the perturbing forces.

In this chapter we will introduce canonical transformations, and solve the two body problem using this technique. This process will introduce new sets of orbital elements which are also canonically conjugate coordinates and momenta, with very, very simple Hamiltonian functions. As we perform this solution, remember that there may be other terms in the Hamiltonian due to the potential energies of perturbing forces. When we include the perturbing forces in later chapters, the canonical equations of motion will lead naturally to perturbation equations of motion for the canonical orbital elements.

2.2 Coordinate Choices

As a review of producing a Hamiltonian function for a dynamical system, take the two dimensional harmonic oscillator of Figure 2.1. At the start, we have the choice of which coordinates to employ in describing the state of this system. Using rectangular coordinates x and y, the kinetic energy is

$$T = \frac{1}{2}m\left(\dot{x}^2 + \dot{y}^2\right) \tag{2.1}$$

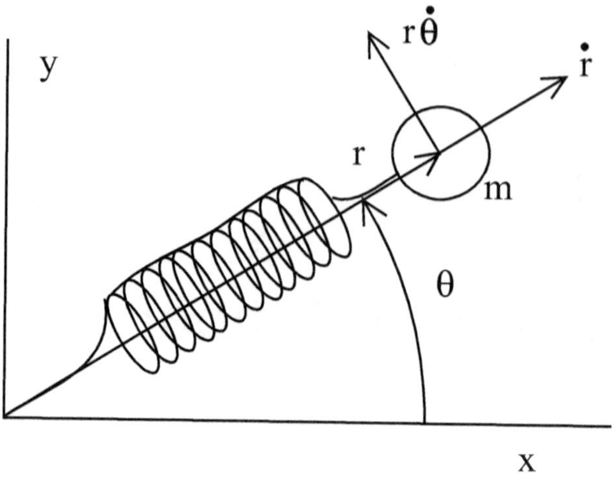

Figure 2.1: The Harmonic Oscillator

while the potential energy from the spring is just

$$V = \frac{1}{2}k\left(x^2 + y^2\right) \tag{2.2}$$

The Lagrangian function is then

$$\begin{aligned} L &= T - V && (2.3) \\ &= \frac{1}{2}m\left(\dot{x}^2 + \dot{y}^2\right) - \frac{1}{2}k\left(x^2 + y^2\right) && (2.4) \end{aligned}$$

At this point, we find out that the generalized momenta

$$p_i = \frac{\partial L}{\partial \dot{q}_i} \tag{2.5}$$

are

$$p_x = m\dot{x}, \quad p_y = m\dot{y} \tag{2.6}$$

or just the linear momentum components. Knowing the momenta, we can form the Hamiltonian function as

$$H = \sum p_i \dot{q}_i - L \tag{2.7}$$

after, of course, *the generalized velocities are eliminated in favor of the generalized momenta.* Since

$$\dot{x} = p_x/m, \quad \dot{y} = p_y/m \tag{2.8}$$

we have

$$\begin{aligned} H &= p_x^2/m + p_y^2/m - \frac{1}{2}m\left((p_x/m)^2 + (p_y/m)^2\right) + \frac{1}{2}k\left(x^2+y^2\right) \\ &= \frac{1}{2m}\left(p_x^2+p_y^2\right) + \frac{1}{2}k\left(x^2+y^2\right) \end{aligned} \tag{2.9}$$

Since time is missing from the Hamiltonian, and

$$\frac{dH}{dt} = \frac{\partial H}{\partial t} \tag{2.10}$$

from Hamiltonian dynamics, we have H itself as an integral of the motion. No other integrals are immediately apparent, since we obtain non – trivial equations of motion for both coordinates and both momenta.

Now, experience a change of heart, and decide that polar coordinates r, θ are a better choice for this problem. Simple kinematics yields the kinetic energy as

$$T = \frac{1}{2}m\left(\dot{r}^2 + r^2\dot{\theta}^2\right) \tag{2.11}$$

while the potential energy expressed in polar coordinates is

$$V = \frac{1}{2}kr^2 \tag{2.12}$$

We then have the Lagrangian function as

$$\begin{aligned} L &= T - V \\ &= \frac{1}{2}m\left(\dot{r}^2 + r^2\dot{\theta}^2\right) - \frac{1}{2}kr^2 \end{aligned} \tag{2.13}$$

At this point we find out that the two momenta

$$p_i = \frac{\partial L}{\partial \dot{q}_i} \tag{2.14}$$

are

$$p_r = m\dot{r}, \quad p_\theta = mr^2\dot{\theta} \tag{2.15}$$

The first is a linear momentum component, since r is a linear coordinate, while the second is an angular momentum component. These relations can easily be solved to yield the generalized velocity components as

$$\dot{r} = p_r/m, \quad \dot{\theta} = p_\theta/mr^2 \tag{2.16}$$

The Hamiltonian can then be found as

$$\begin{aligned} H &= \sum p_i\dot{q}_i - L \\ &= p_r^2/m + p_\theta^2/mr^2 - \frac{1}{2}m\left((p_r/m)^2 + (p_\theta/mr^2)^2\right) + \frac{1}{2}kr^2 \\ &= \frac{1}{2m}\left(p_r^2 + \frac{1}{r^2}p_\theta^2\right) + \frac{1}{2}kr^2 \end{aligned} \tag{2.17}$$

Again, time is missing, so the Hamiltonian function, (which is still the total energy) is a constant of the motion. However, now we have a second integral of the motion, since the coordinate θ is cyclic. This means that the total angular momentum p_θ is also constant.

Now, the total angular momentum was also conserved when we formulated the problem using rectangular coordinates. Our choice of coordinates obscured this fact. Only if you stare at the figure and realize that the problem possess rotational symmetry, or remembered that this is a central force problem, would you suspect that angular momentum is conserved. (All of the known classical integrals of motion are associated with coordinate symmetries). The choice of coordinates can influence how much we learn about the problem when we set it up.

However, when we had our change of heart, it was necessary to begin the formulation of the problem all over again from the beginning. In Hamiltonian dynamics we have free choice of the coordinates, so long as they describe the configuration of the system, but we do *not* have free choice of the generalized momenta variables. We find out what the momenta are only after we have the Lagrangian in our possession, and even then it is not a choice: equations (2.5) or (2.15) *dictate* to us what the momenta must be.

The two Hamiltonian functions (2.9) and (2.17) do represent the same system, and both are the total energy. It must be possible to "jump" from the rectangular coordinate form the the polar coordinate form *without having to start all over again from the beginning.* In doing this, we will have to carefully preserve the structure of Hamiltonian dynamics: we have free choice of only half of the variables, and Hamilton's equations must be obeyed. This change of variables is performed with a canonical transformation. The larger question of how to choose variables to obtain the "best" Hamiltonian is one we will take up shortly.

2.3 Canonical Transformations

A set of coordinates q_i and momenta p_i are canonical if they satisfy Hamilton's equations:

$$\dot{q}_i = \frac{\partial H}{\partial p_i}, \quad \dot{p}_i = -\frac{\partial H}{\partial q_i} \tag{2.18}$$

where $H(q_i, p_i, t)$ is the Hamiltonian function for a particular dynamical system. We have seen how to obtain the Hamiltonian for a particular dynamical system. This process involves choosing the generalized coordinates q_i, and obtaining the Lagrangian function. Once the Lagrangian is obtained, the generalized momenta are defined by

$$p_i = \frac{\partial L}{\partial \dot{q}_i} \tag{2.19}$$

These are linear relations between the p_i and the generalized velocities q_i, which may be inverted to find the velocities in terms of the momenta. Then, the

Hamiltonian is defined as

$$H = \sum_{i=1}^{N} p_i \dot{q}_i - L \tag{2.20}$$

after the $\dot{q}_i$ are eliminated in favor of the momenta p_i.

However, having found the Hamiltonian, we have no freedom to change coordinates. For every choice of generalized coordinates q_i, the conjugate momenta p_i are determined for us by (2.19). Thus, we cannot simply change the definitions of the coordinates and momenta in the Hamiltonian at will, since the structure of Hamilton's equations will be destroyed. But the choice of coordinates can make or break a problem, governing the simplicity of the equations of motion, the number of constants of the motion found, and the possibility of solving the system in closed form. At the moment, if we chose another set of coordinates, we have no choice but to set up the entire dynamical system again, from the beginning, in order to find the Hamiltonian. In this section we will search for a general technique to change coordinates and momenta while still preserving the structure of Hamilton's equations.

Suppose we have a new set of canonical coordinates Q_i and momenta P_i, given as functions of the old coordinates and momenta:

$$Q_i = Q_i(q_i, p_i, t) \tag{2.21}$$

$$P_i = P_i(q_i, p_i, t) \tag{2.22}$$

Also, there will be a new Hamiltonian function $K = K(Q_i, P_i, t)$ and Hamilton's equations will be obeyed:

$$\dot{Q}_i = \frac{\partial K}{\partial P_i}, \quad \dot{P}_i = -\frac{\partial K}{\partial Q_i} \tag{2.23}$$

We wish that the old variables (q_i, p_i) and the new variables (Q_i, P_i) describe the same dynamical system.

Return to Hamilton's Principle

$$\delta \int L dt = 0 \tag{2.24}$$

or, since

$$H = \sum_{i=1}^{N} p_i \dot{q}_i - L \tag{2.25}$$

we must have

$$\delta \int \left(\sum_{i=1}^{N} p_i \dot{q}_i - H \right) dt = 0 \tag{2.26}$$

for the old set of variables, and

$$\delta \int \left(\sum_{i=1}^{N} P_i \dot{Q}_i - K \right) dt = 0 \tag{2.27}$$

for the new variables. The variation of both integrals equals zero, and the integrands do represent the same dynamical system, but it does not follow that the *integrands* themselves are equal.

However, they do differ by the total time derivative of an arbitrary function F:

$$\delta \int \Big(\sum p_i \dot{q}_i - H(p_i, q_i, t) - \sum P_i \dot{Q}_i + K(Q_i, P_i, t) - \frac{dF}{dt} \Big) \, dt = 0 \tag{2.28}$$

This is true since

$$\delta \int_{t_1}^{t_2} \frac{dF}{dt} dt = \delta(F(t_2) - F(t_1)) \tag{2.29}$$

and the virtual variation must vanish at the end times. The function F is called the *generating function* for the transformation.

Now, F could be a function of all 4N+1 variables q_i, p_i, Q_i, P_i, t. However, we have the 2N equations (2.22) linking the old and new variables. So, only 2N+1 of the 4N+1 variables are independent. There are four basic forms for F:

$$\begin{aligned} &F_1(q, Q, t), \quad F_2(q, P, t) \\ &F_3(p, Q, t), \quad F_4(p, P, t) \end{aligned}$$

Taking the first choice, the variables q_i and Q_i are assumed independent. The time derivative of F becomes

$$\frac{dF_1}{dt} = \sum \frac{\partial F_1}{\partial q_i} \dot{q}_i + \sum \frac{\partial F_1}{\partial Q_i} \dot{Q}_i + \frac{\partial F_1}{\partial t} \tag{2.30}$$

Substituting into equation (2.28), and equating the integrand to zero, we find

$$\begin{aligned} \sum p_i \dot{q}_i &- H - \sum P_i \dot{Q}_i + K - \sum \frac{\partial F_1}{\partial q_i} \dot{q}_i \\ &- \sum \frac{\partial F_1}{\partial Q_i} \dot{Q}_i - \frac{\partial F_1}{\partial t} = 0 \end{aligned} \tag{2.31}$$

As the variables Q_i, q_i and t are independent, we can equate the coefficients of $\dot{q}_i$, $\dot{Q}_i$, to zero, obtaining

$$p_i = \frac{\partial F_1}{\partial q_i} \tag{2.32}$$

$$P_i = -\frac{\partial F_1}{\partial Q_i} \tag{2.33}$$

The remainder of the expression yields

$$K(Q, P, t) = H(q, p, t) + \frac{\partial F_1}{\partial t} \tag{2.34}$$

Table I.
Canonical Transform Relations

$F_1(q,Q,t)$	$F_2(q,P,t)$	$F_3(p,Q,t)$	$F_4(p,P,t)$
$p_i = \frac{\partial F_1}{\partial q_i}$	$p_i = \frac{\partial F_2}{\partial q_i}$	$q_i = -\frac{\partial F_3}{\partial p_i}$	$q_i = -\frac{\partial F_4}{\partial p_i}$
$P_i = -\frac{\partial F_1}{\partial Q_i}$	$Q_i = \frac{\partial F_2}{\partial P_i}$	$P_i = -\frac{\partial F_3}{\partial Q_i}$	$Q_i = \frac{\partial F_4}{\partial P_i}$

So, knowing the generating function $F_1(q,Q,t)$ enables one to complete the transformation by finding p_i, P_i, and the new Hamiltonian K.

There are three other possibilities for the generating function. These are defined in terms of the first form as:

$$F_1 = F_2 - \sum P_i Q_i \tag{2.35}$$

$$F_1 = F_3 + \sum p_i q_i \tag{2.36}$$

$$F_1 = F_4 - \sum P_i Q_i + \sum p_i q_i \tag{2.37}$$

Following the second form, the total derivative of the generating function becomes

$$\begin{aligned} \frac{dF_1}{dt} &= \sum \frac{\partial F_2}{\partial q_i}\dot{q}_i + \sum \frac{\partial F_2}{\partial P_i}\dot{P}_i + \frac{\partial F_2}{\partial t} \\ &- \sum P_i \dot{Q}_i - \sum \dot{P}_i Q_i \end{aligned} \tag{2.38}$$

When substituted into Hamilton's principle, this gives

$$\begin{aligned} &\sum p_i \dot{q}_i - H + \sum \dot{P}_i Q_i + K \\ - &\sum \frac{\partial F_2}{\partial q_i}\dot{q}_i - \sum \frac{\partial F_2}{\partial P_i}\dot{P}_i - \frac{\partial F_2}{\partial t} = 0 \end{aligned} \tag{2.39}$$

Equating coefficients of independent variables yields the transformation relations

$$p_i = \frac{\partial F_2}{\partial q_i}, \quad Q_i = \frac{\partial F_2}{\partial P_i}, \quad K = H + \frac{\partial F_2}{\partial t} \tag{2.40}$$

The remaining two possible forms of the generating function are similar in the derivation of the transformation relations. In summary, the complete set of transform relations for canonical transformations are summarized in Table I. For all cases, the new Hamiltonian function is given by

$$K = H + \frac{\partial F}{\partial t} \tag{2.41}$$

after the old variables q_i, p_i have been eliminated in favor of the new variables Q_i, P_i.

2.4 The Harmonic Oscillator

As an example, consider the two dimensional harmonic oscillator we began studying in section 2.2. If the generalized coordinates are x and y, the generalized momenta are p_x and p_y, the linear momentum components, and the Hamiltonian function is:

$$H = \frac{1}{2m}\left(p_x^2 + p_y^2\right) + \frac{1}{2}k\left(x^2 + y^2\right) \tag{2.42}$$

In this form, the only obvious constant of the motion is the Hamiltonian itself, since (2.42) does not explicitly depend on time. The equations of motion are linear, constant coefficient differential equations, and can be solved by elementary techniques. However, suppose we wished to find the equations of motion in another set of coordinates.

One obvious alternative is polar coordinates r, θ. The new coordinates r and θ can be written in terms of the old coordinates as

$$r = \left(x^2 + y^2\right)^{1/2} \tag{2.43}$$

$$\theta = \tan^{-1}\frac{y}{x} \tag{2.44}$$

In this case, we know what the desired new coordinates Q are given in terms of the old coordinates $q = x, y$. To complete the transformation, we need to find the generating function for this transformation.

The relations (2.43) and (2.44) can be interpreted as one half of the transformation laws given in Table I of the previous section. Only the forms F_2 and F_4 are possibilities, since only these two forms have explicit relations yielding the new coordinates. Also, equations (2.43) and (2.44) give the new coordinates in terms of the old coordinates. This eliminates F_4, since this form of the generating function is not a function of the old coordinates, and F_2 is the correct choice. We can then combine the transform law for Q from Table I with the definitions of the new coordinates to give

$$\frac{\partial F_2}{\partial P_r} = \left(x^2 + y^2\right)^{1/2} \tag{2.45}$$

$$\frac{\partial F_2}{\partial P_\theta} = \tan^{-1}\frac{y}{x} \tag{2.46}$$

These are two partial differential equations for the generating function F_2. One possible solution is

$$F_2 = P_r\left(x^2+y^2\right)^{1/2} + P_\theta \tan^{-1}\frac{y}{x} \tag{2.47}$$

The most general solution for F will differ from this by only an arbitrary function of time alone. We do not wish to risk the introduction of time into the new Hamiltonian, since this will cost us an integral of the motion. So, let us continue with the above form.

The other half of the transform laws for F_2 then give the old momenta in terms of the new momenta:

$$p_x = \frac{\partial F_2}{\partial x} = \frac{xP_r}{r} - \frac{yP_\theta}{r^2} \tag{2.48}$$

$$p_y = \frac{\partial F_2}{\partial y} = \frac{yP_r}{r} + \frac{xP_\theta}{r^2} \tag{2.49}$$

The final transformation law reduces to the statement that $K = H$, since $\partial F_2/\partial t = 0$. However, the old coordinates and momenta must be eliminated from K, since K is a function only of the new variables. Obviously $x^2+y^2 = r^2$, and working out the kinetic energy term

$$p_x^2 + p_y^2 = P_r^2 + \frac{1}{r^2}P_\theta^2 \tag{2.50}$$

Then, the new Hamiltonian takes the form:

$$K = \frac{1}{2m}\left(P_r^2 + \frac{1}{r^2}P_\theta^2\right) + \frac{1}{2}kr^2 \tag{2.51}$$

This form possesses an additional constant of the motion. Time is still missing from the new Hamiltonian, so it is still a constant of the motion. However, the new coordinate θ is also cyclic in the new Hamiltonian, so the new momentum P_θ is also conserved. This new momentum represents the total angular momentum of the mass about the origin.

As a final check, the new Hamiltonian (2.51) is exactly what we obtained when we set up the problem from the beginning in polar coordinates. By the use of a canonical transformation we have obtained this result without beginning again.

2.5 Hamilton – Jacobi Theory, I

The last section shows, yet again, that the "correct" choice of coordinates can aid the solution of a dynamical problem. We can ask the question, "what is the simplest form for the equations of motion?". Hamilton's equations of motion in the new variables are

$$\dot{Q}_i = \frac{\partial K}{\partial P_i}, \quad \dot{P}_i = -\frac{\partial K}{\partial Q_i} \tag{2.52}$$

Obviously, the more coordinates that are missing in the Hamiltonian, the more conjugate momenta are constant. Also, if any momenta are absent in the new Hamiltonian, then their conjugate coordinates are also constant. In either case, the relevant equation of motion becomes trivial to solve. Now, the best possible situation would be the case when the new Hamiltonian does not contain any new coordinates or momenta. If the new Hamiltonian is identically zero, then all coordinates and momenta are constants of the motion. Hamilton – Jacobi theory undertakes to solve a dynamical system by a transformation to a new set of coordinates and momenta, all of which are constant.

We can extract the transform relation for the new Hamiltonian from Table I, and explicitly requiring that the new Hamiltonian vanishes yields

$$H(q_i, p_i, t) + \frac{\partial F}{\partial t} = 0 \tag{2.53}$$

This is called the *Hamilton – Jacobi equation.* We have not yet specified the form of the generating function we are using. Since (2.53) is a function of the old coordinates and momenta, we can use any form that contains either of these quantities, and provides a relation for the other. The standard choice, however, is F_2. When this is substituted into (2.53), using the transform relation $p_i = \partial F_2/\partial q_i$, we obtain

$$H\left(q_i, \frac{\partial F_2}{\partial q_i}, t\right) + \frac{\partial F_2}{\partial t} = 0 \tag{2.54}$$

This is a partial differential equation for F_2, termed in this application *Hamilton's Principal Function,* and usually denoted $F_2 = S$. Its solution transforms a given dynamical system to a new set of coordinates and momenta all of which are constant. Hamilton was the first to derive this partial differential equation, and show that its solutions produced the actual solution to dynamical systems. Jacobi's contribution was to show that any solution to this partial differential equation would have the desired property. We do not need to seek the most general solution to (2.54), since any old solution with a full set of arbitrary constants will do.

This is a startling concept when first encountered. The idea that the solution to any dynamical system involves a new phase space in which all coordinates and momenta are constant is not an obvious one. It was much studied in the 50 years following Hamilton's and Jacobi's contributions, and was the insight that led Einstein to pose "straight line motion in a curved space" as one of the fundamental concepts of general relativity.

Since the new coordinates and momenta will all be constant, it is the transformation laws themselves which contain the actual solution. They will give the old coordinates and momenta in terms of the new coordinates and momenta (all constant) and the time. With expressions for the old coordinates and momenta in terms of 2N arbitrary constants and time, we have the actual solution in our possession. Rather than continue with the formal development, it is best to pause for some examples.

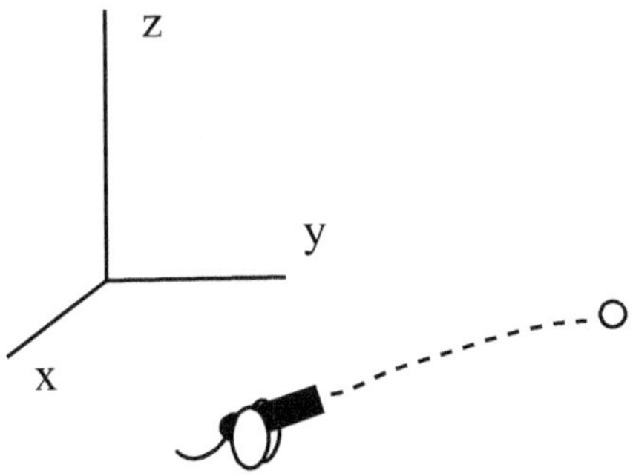

Figure 2.2: The Cannon Ball Problem

2.6 The Cannonball Problem

One of the old war – horses of dynamics, shown in Figure 2.2, is the problem of a projectile in free flight above the surface of the earth. First solved by Galileo, we will now solve it by our new technique of Hamilton – Jacobi theory. In rectangular coordinates, with the z axis up, the system Hamiltonian is

$$H = \frac{1}{2m}\left(p_x^2 + p_y^2 + p_z^2\right) + mgz \tag{2.55}$$

To obtain the Hamilton – Jacobi equation, we replace each momenta p_i with the partial of the generating function $\partial S/\partial q_i$, and add $\partial S/\partial t$, setting the result to zero. The Hamilton – Jacobi partial differential equation becomes

$$\frac{1}{2m}\left(\left(\frac{\partial S}{\partial x}\right)^2 + \left(\frac{\partial S}{\partial y}\right)^2 + \left(\frac{\partial S}{\partial z}\right)^2\right) + mgz + \frac{\partial S}{\partial t} = 0 \tag{2.56}$$

Now, remember Jacobi's result: we do not need the most general solution to this equation. Any solution containing 3 new arbitrary constants is sufficient for our purposes. The Hamilton – Jacobi equation above can be solved by the technique of separation of variables. If we assume S has the form

$$S = S_1(x) + S_2(y) + S_3(z) + S_4(t) \tag{2.57}$$

and insert this into (2.56), we have

$$\frac{1}{2m}\left(\left(\frac{dS_1}{dx}\right)^2 + \left(\frac{dS_2}{dy}\right)^2 + \left(\frac{dS_3}{dz}\right)^2\right) + mgz = -\frac{dS_4}{dt} \tag{2.58}$$

The partial derivatives have become total derivatives, since each of the component functions in (2.58) is a function of only one variable. In the above expression, the left side is a function of x, y, and z; but not the time t. The right side is a function of time, but not x, y, or z. This can only be true for any values of x, y, z, and t if both sides are equal to a constant.

Now, the S function is an F_2 generating function, and depends on the old coordinates q_i, and the new momenta P_i. However, the new momenta P_i are all constants of the motion, since the new Hamiltonian is zero. We are free to interpret the new constant as one of the new momenta, say P_1. In fact, since the left side is still the system Hamiltonian in another guise, this new momentum P_1 must simply be the total (constant) energy of the projectile. Equation (2.58) can be separated into the two statements:

$$\frac{dS_4}{dt} = -P_1 \tag{2.59}$$

$$\frac{1}{2m}\left(\left(\frac{dS_1}{dx}\right)^2 + \left(\frac{dS_2}{dy}\right)^2 + \left(\frac{dS_3}{dz}\right)^2\right) + mgz = P_1 \tag{2.60}$$

The first separated equation immediately integrates to give

$$S_4(t) = -P_1 t \tag{2.61}$$

Continuing with this process, the second equation can be rearranged to give

$$\frac{1}{2m}\left(\frac{dS_1}{dx}\right)^2 = -\frac{1}{2m}\left(\left(\frac{dS_2}{dy}\right)^2 + \left(\frac{dS_3}{dz}\right)^2\right) - mgz + P_1 \tag{2.62}$$

Again, the left side is a function only of x, while the right side contains all dependence on y and z. This can only be true for any values of x, y, and z if both sides equal the same constant. This constant is our second momentum P_2. Separating the two sides gives

$$\frac{1}{2m}\left(\frac{dS_1}{dx}\right)^2 = P_2 \tag{2.63}$$

$$-\frac{1}{2m}\left(\left(\frac{dS_2}{dy}\right)^2 + \left(\frac{dS_3}{dz}\right)^2\right) - mgz + P_1 = P_2 \tag{2.64}$$

The first line integrates to give us the second piece of Hamilton's Principal Function

$$S_1(x) = \sqrt{2mP_2}\, x \tag{2.65}$$

The final separation yields the two equations

$$\frac{1}{2m}\left(\frac{dS_2}{dy}\right)^2 = P_3 \tag{2.66}$$

$$-\frac{1}{2m}\left(\frac{dS_3}{dz}\right)^2 - mgz + P_1 - P_2 = P_3 \tag{2.67}$$

and introduces the third constant momentum P_3. Both equations easily integrate to give

$$S_2(y) = \sqrt{2mP_3}\, y \tag{2.68}$$

$$S_3(z) = -\frac{1}{3m^2 g}\left(2m\left(P_1 - P_2 - P_3 - mgz\right)\right)^{3/2} \tag{2.69}$$

So, we now have in our possession the complete generating function:

$$\begin{aligned} S &= \sqrt{2mP_2}\, x + \sqrt{2mP_3}\, y - P_1 t \\ &- \frac{1}{3m^2 g}\left(2m\left(P_1 - P_2 - P_3 - mgz\right)\right)^{3/2} \end{aligned} \tag{2.70}$$

Remember that S is an F_2 generating function. The new constant coordinates are obtained from the relation

$$Q_i = \frac{\partial S}{\partial P_i} \tag{2.71}$$

from Table I. These give

$$Q_1 = -t - \frac{1}{mg}\sqrt{2m\left(P_1 - P_2 - P_3 - mgz\right)} \tag{2.72}$$

$$Q_2 = \sqrt{m/2P_2}x + \frac{1}{mg}\sqrt{2m\left(P_1 - P_2 - P_3 - mgz\right)} \tag{2.73}$$

$$Q_3 = \sqrt{m/2P_3}y + \frac{1}{mg}\sqrt{2m\left(P_1 - P_2 - P_3 - mgz\right)} \tag{2.74}$$

These expressions could be used to evaluate the constant Q_i at some initial time. However, we are much more interested in the inverse of these expressions.

Solving the first for z gives

$$z = -\frac{1}{2}g\left(t + Q_1\right)^2 + \frac{1}{mg}\left(P_1 - P_2 - P_3\right) \tag{2.75}$$

This is the familiar quadratic expression in time, giving the parabolic shape to the trajectory. Notice that $-Q_1$ must be a time, and in fact is the time the projectile reaches the top of its trajectory. Adding the Q_1 and Q_2 equations produces a result easily solved for x:

$$x = \sqrt{2P_2/m}\left(Q_1 + Q_2 + t\right) \tag{2.76}$$

Similarly, adding Q_1 and Q_3 gives

$$y = \sqrt{2P_3/m}\left(Q_1 + Q_3 + t\right) \tag{2.77}$$

These are linear expressions in time, giving the familiar result that the projectile moves across the surface of the earth in a straight line at constant speed. Again, the two new constant coordinates Q_2 and Q_3 must have units of time.

The other half of the F_2 transform relations give the old momenta in terms of the new momenta:

$$p_x = \sqrt{2mP_2} \tag{2.78}$$

$$p_y = \sqrt{2mP_3} \tag{2.79}$$

$$p_z = \sqrt{2m\left(P_1 - P_2 - P_3 - mgz\right)} \tag{2.80}$$

The new momenta P_2 and P_3 are the x and y "components" of the kinetic energy of the cannonball, while the first momentum P_1 is, by equation (2.60), the total energy of the projectile.

Solving the cannonball problem with Hamilton – Jacobi theory is a slight case of overkill. However, Hamilton – Jacobi theory is not all–powerful, since only systems solvable by the usual techniques will also yield to this new approach. However, we will find that Hamilton – Jacobi theory is the natural first step in attacking a problem with perturbation techniques.

2.7 The Planar Harmonic Oscillator

As another example of the Hamilton – Jacobi method, consider the two dimensional harmonic oscillator in rectangular coordinates. Its' Hamiltonian is:

$$H = \frac{1}{2m}\left(p_x^2 + p_y^2\right) + \frac{k}{2}\left(x^2 + y^2\right) \tag{2.81}$$

The Hamilton – Jacobi equation is obtained by replacing the momenta p_i with $\partial S/\partial q_i$, and is

$$\frac{1}{2m}\left(\left(\frac{\partial S}{\partial x}\right)^2 + \left(\frac{\partial S}{\partial y}\right)^2\right) + \frac{k}{2}\left(x^2 + y^2\right) + \frac{\partial S}{\partial t} = 0 \tag{2.82}$$

Making the usual assumption that Hamilton's principal function is separable as

$$S = S_x(x) + S_y(y) + S_t(t) \tag{2.83}$$

the Hamilton – Jacobi equation becomes, after a slight rearrangement:

$$\frac{1}{2m}\left(\left(\frac{dS_x}{dx}\right)^2 + \left(\frac{dS_y}{dy}\right)^2\right) + \frac{k}{2}\left(x^2 + y^2\right) = -\frac{dS_t}{dt} = P_1 \tag{2.84}$$

The first separation constant, P_1, is obviously the total energy of the oscillator.

This immediately gives the first piece of the generating function:

$$S_t = -P_1 t \tag{2.85}$$

while the remaining portion of the Hamilton–Jacobi equation can be separated again to give

$$\frac{1}{2m}\left(\frac{dS_x}{dx}\right)^2 + \frac{k}{2}x^2 = P_1 - \frac{1}{2m}\left(\frac{dS_y}{dy}\right)^2 - \frac{k}{2}y^2 = P_2 \tag{2.86}$$

The second separation constant P_2 can be identified as the portion of the total energy contained within the x coordinate motion. This last separation immediately gives the generating function in terms of two indefinite integrals. Assembling the entire generating function, we have

$$\begin{aligned} S &= -P_1 t + \sqrt{2m}\int \sqrt{P_2 - kx^2/2}\, dx \\ &+ \sqrt{2m}\int \sqrt{P_1 - P_2 - ky^2/2}\, dy \end{aligned} \tag{2.87}$$

The integrals *could* be carried out with the aid of a very good integral table, but notice that we have no real use for the generating function itself. Rather, we are interested in the coordinate transformation.

The new coordinates Q_1 and Q_2 are given by

$$\begin{aligned} Q_1 &= \frac{\partial S}{\partial P_1} = -t + \sqrt{\frac{m}{k}}\int \left(\frac{P_1 - P_2}{k/2} - y^2\right)^{-1/2} dy \\ &= -t + \sqrt{\frac{m}{k}}\sin^{-1}\left(\frac{y}{\sqrt{2\,(P_1 - P_2)\,/k}}\right) \end{aligned} \tag{2.88}$$

and

$$\begin{aligned} Q_2 &= \frac{\partial S}{\partial P_2} = -\sqrt{\frac{m}{k}}\int \left(\frac{P_1 - P_2}{k/2} - y^2\right)^{-1/2} dy \\ &+ \sqrt{\frac{m}{k}}\int \left(\frac{P_2}{k/2} - x^2\right)^{-1/2} dx \\ &= -\sqrt{\frac{m}{k}}\sin^{-1}\left(\frac{y}{\sqrt{2\,(P_1 - P_2)\,/k}}\right) \\ &+ \sqrt{\frac{m}{k}}\sin^{-1}\left(\frac{x}{\sqrt{2P_2/k}}\right) \end{aligned} \tag{2.89}$$

These equations are then easily solved to give the behavior of the physical coordinates

$$x = \sqrt{\frac{2}{k}P_2}\, \sin\left(\sqrt{\frac{k}{m}}\,(t + Q_1 + Q_2)\right) \tag{2.90}$$

$$y = \sqrt{\frac{2}{k}\,(P_1 - P_2)}\, \sin\left(\sqrt{\frac{k}{m}}\,(t + Q_1)\right) \tag{2.91}$$

The solution for the physical coordinates x and y show the expected sinusoidal behavior, with the expected frequency $\omega = \sqrt{k/m}$. The amplitudes of the x and y motions are proportional to the square roots of the oscillation energies in each direction, while the Q_i have units of time, and are the phase variables needed to satisfy arbitrary initial conditions.

The other relations for the F_2 generating function will yield the old momenta as

$$p_i = \frac{\partial S}{\partial q_i} \tag{2.92}$$

However, note that this would require us to do two more integrals. It is easier to obtain the old momenta by remembering their definitions in terms of the generalized velocities

$$p_x = m\dot{x} = \sqrt{2mP_2}\,\cos\left(\sqrt{\frac{k}{m}}\,(t + Q_1 + Q_2)\right) \tag{2.93}$$

$$p_y = m\dot{y} = \sqrt{2m\,(P_1 - P_2)}\,\cos\left(\sqrt{\frac{k}{m}}\,(t + Q_1)\right) \tag{2.94}$$

Together with equations (2.90) and (2.91), these constitute the complete solution to the two dimensional harmonic oscillator.

The new Hamiltonian K can now be calculated as a check. It is given by

$$K = H + \frac{\partial S}{\partial t} \tag{2.95}$$

after the old coordinates and momenta are eliminated in favor of the new variables. Substituting in (2.95), we have

$$\begin{aligned} K &= \frac{1}{2m}\left(2mP_2\cos^2\alpha + 2m\,(P_1 - P_2)\cos^2\beta\right) \\ &+ \frac{k}{2}\left(\frac{2}{k}P_2\sin^2\alpha + \frac{2}{k}\,(P_1 - P_2)\sin^2\beta\right) - P_1 \end{aligned} \tag{2.96}$$

where the angles are abbreviated as

$$\alpha = \sqrt{\frac{k}{m}}\,(t + Q_1 + Q_2) \tag{2.97}$$

and

$$\beta = \sqrt{\frac{k}{m}}\,(t + Q_1) \tag{2.98}$$

Combining terms, this reduces to

$$K = P_2 + P_1 - P_2 - P_1 = 0 \tag{2.99}$$

So, as advertised, the new Hamiltonian is identically zero, and all new coordinates and momenta are constant.

The solution appears as equations (2.90), (2.91), (2.93), and (2.94) as the coordinate transformation laws. The new variables P_i, Q_i are all constant, as shown by the vanishing of the new Hamiltonian K. As constants of the motion, they can be evaluated from equations (2.84), (2.86), (2.88), and (2.89) in terms of initial conditions. Knowing their values for a particular trajectory, we have completed the solution.

2.8 Hamilton – Jacobi Theory, II

A very common case in Hamilton – Jacobi theory occurs wherever the Hamiltonian is not a function of time, and is therefore itself a constant of the motion. In this case, the Hamilton–Jacobi equation

$$H\left(q_i, \frac{\partial S}{\partial q_i}\right) + \frac{\partial S}{\partial t} = 0 \tag{2.100}$$

can always be taken through at least one separation step. Write Hamilton's principal function S as

$$S(q_i, P_i, t) = W(q_i, P_i) + S_t(t) \tag{2.101}$$

Then, the Hamilton – Jacobi equation becomes

$$H\left(q_i, \frac{\partial W}{\partial q_i}\right) + \frac{dS_t}{dt} = 0 \tag{2.102}$$

This immediately separates to give

$$\frac{dS_t}{dt} = -P_1 \tag{2.103}$$

$$H\left(q_i, \frac{\partial W}{\partial q_i}\right) = P_1 \tag{2.104}$$

The first relation integrates to give

$$S_t(t) = -P_1 t \tag{2.105}$$

while the second relation is the new form of the Hamilton – Jacobi partial differential equation. The first separation momentum P_1 will be the total energy of the system in those (frequent) cases when the Hamiltonian itself is the total energy. The function $W(q, P)$ is called *Hamilton's characteristic function.* Equation (2.101) relates Hamilton's principal function to Hamilton's characteristic function. Starting with the characteristic function W saves one separation step when the system Hamiltonian is conserved.

2.9 The Two Body Problem

Let us take Hamilton – Jacobi theory, and attempt to solve the two body problem with it. As shown in Figure 2.3, we will start with spherical polar coordinates r, θ, ϕ. The potential energy per unit mass is

$$V = -\frac{\mu}{r} \tag{2.106}$$

where μ is the gravitational parameter. The velocity vector of the satellite is

$$\mathbf{v} = \dot{r}\mathbf{e}_r + r\dot{\theta}\mathbf{e}_\theta + r\sin\theta\dot{\phi}\mathbf{e}_\phi \tag{2.107}$$

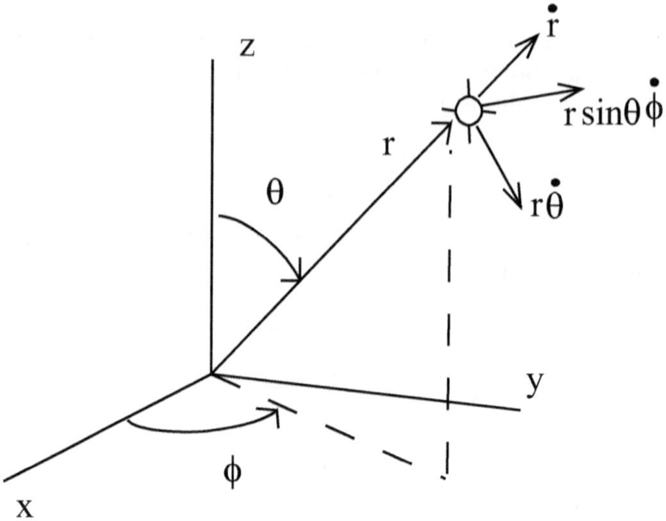

Figure 2.3: Coordinates for the Two Body Problem

so the Lagrangian per unit mass is

$$L = \frac{1}{2}\left(\dot{r}^2 + r^2\dot{\theta}^2 + r^2\sin^2\theta\dot{\phi}^2\right) + \frac{\mu}{r} \tag{2.108}$$

Notice that ϕ is cyclic, so the momentum p_ϕ is already a constant of the motion. Also time does not appear, so the Hamiltonian will also be constant. The canonical momenta are

$$p_r = \frac{\partial L}{\partial \dot{r}} = \dot{r} \tag{2.109}$$

$$p_\theta = \frac{\partial L}{\partial \dot{\theta}} = r^2\dot{\theta} \tag{2.110}$$

$$p_\phi = \frac{\partial L}{\partial \dot{\phi}} = r^2\sin^2\theta\dot{\phi} \tag{2.111}$$

and we recognize p_ϕ as the z component of the total angular momentum. So, the Hamiltonian function is:

$$H = \frac{1}{2}\left(p_r^2 + \frac{1}{r^2}p_\theta^2 + \frac{1}{r^2\sin^2\theta}p_\phi^2\right) - \frac{\mu}{r} \tag{2.112}$$

Since the Hamiltonian is not a function of time, we may begin with the

Hamilton – Jacobi equation for Hamilton's characteristic function W:

$$\frac{1}{2}\left(\left(\frac{\partial W}{\partial r}\right)^2 + \frac{1}{r^2}\left(\frac{\partial W}{\partial \theta}\right)^2 + \frac{1}{r^2 \sin^2\theta}\left(\frac{\partial W}{\partial \phi}\right)^2\right) - \frac{\mu}{r} = P_E \tag{2.113}$$

where the first new momentum P_E is the constant total energy of the orbit. In spherical coordinates, this equation is separable.

Assume that the generating function takes the form

$$W = W_r(r) + W_\theta(\theta) + W_\phi(\phi) \tag{2.114}$$

Then the partial derivatives in (2.113) become total derivatives. Also, the only ϕ dependence is contained within the term $\partial W/\partial\phi$. We could solve for this term, and make the usual argument that one side depends only on ϕ, while the other side depends on r and θ. This will introduce the second constant momentum P_ϕ, and produce the relation

$$\frac{dW_\phi}{d\phi} = P_\phi \tag{2.115}$$

Since $dW_\phi/d\phi$ is just p_ϕ, we recognize that $P_\phi = p_\phi$.

Now, it is not necessary to actually separate the equation to make this argument. Merely assume that we have done the separation, introduced the separation constant, and then reassembled the pieces to look like before. The other side, depending on r and θ, will become

$$\frac{1}{2}\left(\left(\frac{dW_r}{dr}\right)^2 + \frac{1}{r^2}\left\{\left(\frac{dW}{d\theta}\right)^2 + \frac{1}{\sin^2\theta}P_\phi^2\right\}\right) - \frac{\mu}{r} = P_E \tag{2.116}$$

Now, the only portion of (2.116) which depends on θ is the portion enclosed within the inner braces. This too can be separated to yield

$$\left(\frac{dW}{d\theta}\right)^2 + \frac{1}{\sin^2\theta}P_\phi^2 = P_3^2 \tag{2.117}$$

This also must be a constant. However, in this case we have chosen to pick the constant to be P_3^2. This is an arbitrary choice, since we know that the new momenta are constant, but we do not know that they must be the actual constants produced by the separation of the Hamilton – Jacobi equation. This introduces our third and last new constant momentum.

Why we have chosen P_3^2 becomes more obvious when the Hamilton – Jacobi equation is reunited:

$$\frac{1}{2}\left(\left(\frac{dW_r}{dr}\right)^2 + \frac{1}{r^2}P_3^2\right) - \frac{\mu}{r} = P_E \tag{2.118}$$

Let us compare this to the two body problem Hamiltonian function in planar r, ψ coordinates, where the plane of reference is chosen to be the orbital plane:

$$H_{flat} = \frac{1}{2}\left(p_r^2 + \frac{1}{r^2}p_\psi^2\right) - \frac{\mu}{r} \tag{2.119}$$

Here, p_ψ is the total angular momentum of the orbit. Comparing (2.118) to (2.119), it is apparent that $P_3 = p_\psi$, so our third momenta is the total angular momentum of the orbit.

So, the separation of the Hamilton - Jacobi equation for the two body problem has introduced three new constant momenta. These are P_E, the total orbital energy, P_3, the total angular momentum of the orbit, and P_ϕ, the z component of the total angular momentum. To complete the solution, we must use the remaining transformation relations for the F_2 generating function. While this process can be carried out in closed form producing the complete two body problem solution, it involves painful use of contour integration in the complex plane. The solution is given in its entirety in either Goldstein or Szebehely. However, we will find that it is easy enough to guess the final results.

The new constant coordinate Q_E conjugate to the energy P_E must be a constant time. It must be related to the time of perigee passage T_o. By analogy with the harmonic oscillator and cannonball problems, it is most convenient to take this as the *negative* of the perigee passage time. So $Q_E = -T_o$, the negative of the time of perigee passage.

The new constant coordinate Q_3 is conjugate to the total angular momentum of the orbit. Thus, Q_3 must be a constant angle which lies within the plane of the orbit. A well known constant angle which lies within the orbital plane is the argument of perigee, ω. So $Q_3 = \omega$, the argument of perigee.

Finally, the last constant coordinate Q_ϕ is conjugate to the z component of angular momentum P_ϕ. So Q_ϕ is a constant angle within the equator plane of the earth. This last coordinate then must be $Q_\phi = \Omega$, the right ascension of the ascending node.

So, finally, we are led to the canonical set of coordinates

Momenta	Coordinate
P_E, energy	$-T_o$, negative of perigee time
P_3, angular momentum	ω, argument of perigee
P_ϕ, z angular momentum	Ω, node

with a new Hamiltonian function $K = 0$. All of these coordinates and momenta are constants for the two body problem.

2.10 Canonical Elements

In the last section we introduced the natural set of canonical elements for the two body problem. These are

Coordinate	Momentum
$-T_o$, - perigee passage time	E, energy
ω, argument of perigee	h, angular momentum
Ω, node	$h\cos i$, z component of angular momentum

This canonical set of variables possesses an identically zero Hamiltonian, so all are constants of the motion. That is, in the absence of extra potential terms in the Hamiltonian arising from perturbing forces, all six variables are constants.

However, this is not the standard set of canonical variables usually used in perturbation theory. The last four in the table above are standard, but the first two are transformed to slightly different quantities. Also, the last four are renamed. This set of canonical variables is called the *Delaunay elements,* and denoted L, G, H, l, g, h. Capitals denote the momenta. Introduce the generating function

$$F_2 = -\frac{\mu^2}{2L^2}(-T_o) + G\omega + H\Omega - \frac{\mu^2}{2L^2}t \tag{2.120}$$

This transformation is an identity transform in the variables $G = h$, $H = h\cos i$, $g = \omega$, and $h = \Omega$. In the energy variables, the F_2 transform laws give

$$P_E = E = \frac{\partial F_2}{\partial(-T_o)} = -\frac{\mu^2}{2L^2} = -\frac{\mu}{2a} \tag{2.121}$$

The last step is performed using the two body expression for the total orbital energy. This gives the new Delaunay momenta $L = \sqrt{\mu a}$. The other F_2 transform law is

$$l = \frac{\partial F_2}{\partial L} = \frac{\mu^2}{L^3}(t - T_o) \tag{2.122}$$

Using the definition of L, this becomes

$$l = \sqrt{\frac{\mu}{a^3}}(t - T_o) = M \tag{2.123}$$

where M is the mean anomaly.

So, the new Delaunay elements are given by

Momenta	Coordinate
$L = \sqrt{\mu a}$	$l = M$
$G = L\sqrt{1-e^2}$	$g = \omega$
$H = G\cos i$	$h = \Omega$

With a new Hamiltonian

$$K = 0 + \frac{\partial F_2}{\partial t} = -\frac{\mu^2}{2L^2} \tag{2.124}$$

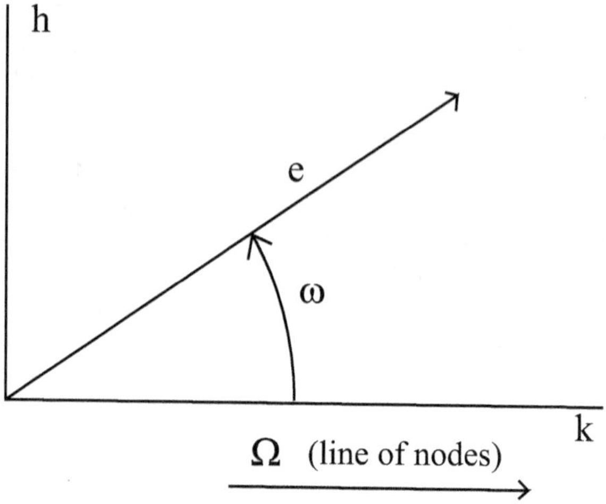

Figure 2.4: The Eccentricity Vector

The momenta G and H are still the total angular momentum, and its z component, respectively. However, the new first momentum L is proportional to the square root of the orbital semimajor axis, while its' conjugate coordinate l is just the mean anomaly. Only the mean anomaly has a non – zero rate, by the Hamiltonian (2.124), and direct calculation will show that the rate is simply $n = \sqrt{\mu/a^3}$, the mean motion.

The Delaunay elements, like the classical elements upon which they are based, have difficulties when the eccentricity $e \to 0$, or when the inclination $i \to 0$. In the first case, the argument of perigee becomes undefined, while in the second case, the node becomes indeterminate. Figure 2.4 shows the eccentricity vector in the orbital plane. As its length e goes to zero, the polar angle ω is not defined. However, this is only a problem if we use polar coordinates. The rectangular coordinates h and k do not experience any difficulties as $e \to 0$. A simple transformation can exchange G and g for h and k. While these variables are not canonical, they are often used to describe low eccentricity orbits. Similarly, the indeterminacy of the node as $i \to 0$ can also be eliminated by a transformation to rectangular rather than polar coordinates. This leads to the classical set of elements:

$$
\begin{aligned}
L &= \sqrt{\mu a} & l &= M \\
h &= e \sin \omega & k &= e \cos \omega \\
p &= \tan i \sin \Omega & q &= \tan i \cos \Omega
\end{aligned}
$$

These are no longer the canonical Delaunay elements, but they have been extensively used for perturbation theory in the solar system, where both inclinations and eccentricities are usually small. Their rates of change must be found by taking time derivatives of their definitions above.

However, the p and q elements above still experience difficulty when the inclination $i \to 90^o$. There are many equatorial earth satellites, but polar orbits are also used for earth satellites. A third set of elements is therefore most often used for earth orbiting objects. Called the *equinoctal elements,* they are given by:

$$\begin{array}{rr} L = \sqrt{\mu a} & \lambda = M + \omega + \Omega \\ h = e\sin(\omega + \Omega) & k = e\cos(\omega + \Omega) \\ \chi = \tan(i/2)\sin\Omega & \psi = \tan(i/2)\cos\Omega \end{array}$$

The equinoctal elements will experience difficulty near $i = 180^o$, or retrograde equatorial orbits. This type of orbit has never been used for an artificial earth satellite, the largest inclination yet launched being less than $i = 130^o$. This is often the standard element set for internal computations in the computer, although classical elements are still used when the computer talks to humans. These are also not a canonical set, so their equations of motion must be found by direct calculation of time derivatives in the above.

By now you have probably noticed the rampant confusion among symbols. The letters h and k are particularly abused, although there are duplications among the other sets as well. It is rare to use more than one or two sets of elements in a calculation, so this reduces confusion somewhat.

2.11 The Extended Hamiltonian

We are used to looking at a Hamiltonian H to see if it contains explicit time dependence. If it does not, then the canonical relation

$$\frac{dH}{dt} = \frac{\partial H}{\partial t} \tag{2.125}$$

states that the Hamiltonian itself is a constant of the motion. However, all canonical relations occur in pairs, and this single equation (2.125) seems to be an exception. It does show that the Hamiltonian is intimately related to the time, almost as if they were canonically conjugate variables.

Define a new momentum $p_{n+1} = -H$, and a new Hamiltonian $\mathcal{H}$ as

$$\mathcal{H}(q_i, p_i, t, p_{n+1}) = H(q_i, p_i, t) + p_{n+1} \tag{2.126}$$

Extreme restraint must be exercised to not "simplify" $\mathcal{H}$ by substituting the definition of p_{n+1}. Numerically $\mathcal{H}$ is identically zero, but the new momentum p_{n+1} and the original Hamiltonian H are to be considered independent variables.

Now, we have no idea what the conjugate coordinate q_{n+1} might be. However, if we write out Hamilton's equations for the new variables, we find

$$\dot{q}_{n+1} = \frac{\partial \mathcal{H}}{\partial p_{n+1}} = 1 \tag{2.127}$$

$$\dot{p}_{n+1} = -\frac{\partial \mathcal{H}}{\partial q_{n+1}} = -\frac{\partial H}{\partial q_{n+1}} \tag{2.128}$$

The first of these, $\dot{q}_{n+1} = 1$ immediately tells us that the new coordinate and the physical time t differ at most by an additive constant. If we choose this constant to be zero, then $q_{n+1} = t$, the time. Then equation (2.128) becomes

$$\dot{p}_{n+1} = -\frac{\partial}{\partial q_{n+1}} H(q_i, p_i, q_{n+1}) \tag{2.129}$$

This is just equation (2.125) rewritten. So, by some slight of hand, time t becomes a canonical coordinate q_{n+1}, conjugate to a momentum p_{n+1} which is (usually) the negative of the system energy. Time and energy are canonically conjugate variables.

However, we have accomplished much more than this. Reducing time and energy to canonical variables makes it possible to do canonical transformations including the variable q_{n+1}. In other words, we can now do changes of *independent variable* by means of canonical transformations. Furthermore, there is now *absolutely nothing* which makes the original Hamiltonian H different from any other canonical variable. Since there is no longer anything special about the status of the system energy, *any other coordinate or momentum can take its place.* Take the original Hamiltonian H

$$H = H(q_i, p_i, q_{n+1}) \tag{2.130}$$

and solve it for one of the momenta, say p_j. Then

$$p_j = p_j(q_i, p_{i,\, i\neq j}, q_{n+1}) = -K(q_i, p_{i,\, i\neq j}, q_{n+1}) \tag{2.131}$$

must also be a Hamiltonian function that describes the same system. The new independent variable must be q_j, and the new Hamiltonian $K = -p_j$ is only constant if K is free of q_j. Hamilton's equations now become

$$\frac{dq_i}{dq_j} = \frac{\partial K}{\partial p_i} \tag{2.132}$$

$$\frac{dp_i}{dq_j} = -\frac{\partial K}{\partial q_i} \tag{2.133}$$

As an example of the use of time transformations, start with the Hamiltonian for the planar two body problem in rectangular coordinates

$$H = \frac{1}{2}\left(p_x^2 + p_y^2\right) - \mu\frac{1}{\sqrt{x^2+y^2}} \tag{2.134}$$

and begin by introducing new coordinates Q_i expressed in terms of the old coordinates by

$$x = Q_1^2 - Q_2^2, \quad y = 2Q_1Q_2 \tag{2.135}$$

Notice that, after some algebra,

$$x^2 + y^2 = \left(Q_1^2 + Q_2^2\right)^2 \tag{2.136}$$

Since we have the old coordinates expressed in terms of the new, the appropriate generating function for this transformation is

$$F_3 = -p_x(Q_1^2 - Q_2^2) - p_y(2Q_1Q_2) \tag{2.137}$$

The other half of the transform laws give the new momenta as

$$P_1 = -\frac{\partial F_3}{\partial Q_1} = 2p_xQ_1 + 2p_yQ_2 \tag{2.138}$$

$$P_2 = -\frac{\partial F_3}{\partial Q_2} = -2p_xQ_2 + 2p_yQ_1 \tag{2.139}$$

Again, after some algebra

$$P_1^2 + P_2^2 = 4(Q_1^2 + Q_2^2)(p_x^2 + p_y^2) \tag{2.140}$$

and substituting, the new Hamiltonian becomes

$$K = \frac{1}{8}\frac{(P_1^2 + P_2^2)}{(Q_1^2 + Q_2^2)} - \frac{\mu}{(Q_1^2 + Q_2^2)} \tag{2.141}$$

Now, following Levi – Civita, convert this to an extended Hamiltonian

$$\mathcal{K} = \frac{1}{8}\frac{(P_1^2 + P_2^2)}{(Q_1^2 + Q_2^2)} - \frac{\mu}{(Q_1^2 + Q_2^2)} + P_3 \tag{2.142}$$

and we prepare to convert the independent variable from the real time t (also Q_3) to a new "pseudotime" τ. The trivial equation of motion $\dot{Q}_3 = 1$ can be made more complicated if we multiply the extended Hamiltonian $\mathcal{K}$ by some function $f(Q_i)$ of the coordinates. If the Hamiltonian were $\mathcal{K}' = \mathcal{K}f(Q_i)$, then the equation of motion for the time would be

$$\frac{dQ_3}{d\tau} = \frac{\partial \mathcal{K}'}{\partial P_3} = f(Q_i) \tag{2.143}$$

and the real time Q_3 and the pseudotime τ are now different. Time transformations are performed by multiplying the extended Hamiltonian by a function of the coordinates. The new independent variable τ can then be made to run faster or slower at different points in space.

Levi – Civita picked

$$f = Q_1^2 + Q_2^2 = r \tag{2.144}$$

the radius vector. As r becomes small, the rate of change of τ increases, effectively "slowing" the orbit near perigee. Conversely, near apogee, when nothing much is happening, the pseudotime τ increases its rate. However, the nicest thing about this time transformation is yet to come.

The new extended Hamiltonian is

$$\mathcal{K}' = f(Q_i)\mathcal{K} = \frac{1}{8}(P_1^2 + P_2^2) - \mu + P_3(Q_1^2 + Q_2^2) \tag{2.145}$$

and the equations of motion are

$$\begin{aligned}
\frac{dQ_1}{d\tau} &= \frac{\partial \mathcal{K}'}{\partial P_1} = \frac{1}{4}P_1 \quad , \quad & \frac{dP_1}{d\tau} &= -\frac{\partial \mathcal{K}'}{\partial Q_1} = -2P_3Q_1 \\
\frac{dQ_2}{d\tau} &= \frac{\partial \mathcal{K}'}{\partial P_2} = \frac{1}{4}P_2 \quad , \quad & \frac{dP_2}{d\tau} &= -\frac{\partial \mathcal{K}'}{\partial Q_2} = -2P_3Q_2 \\
\frac{dQ_3}{d\tau} &= \frac{\partial \mathcal{K}'}{\partial P_3} = Q_1^2 + Q_2^2 \quad , \quad & \frac{dP_3}{d\tau} &= -\frac{\partial \mathcal{K}'}{\partial Q_3} = 0
\end{aligned} \tag{2.146}$$

Since P_3, the negative of the total energy, is constant, this is a harmonic oscillator if the energy is negative, and an exponential system if the energy is positive. With the right coordinates and time variable, the two body problem and the harmonic oscillator are the *same*. The Q_3 equation of motion easily integrates to give the analogue of Kepler's equation for this system.

After discovering this for the planar case, Levi – Civita spent the rest of his life trying to make the three dimensional case work. It can be made to work, but it requires transforming the three dimensional two body problem into a *four* dimensional space. This was discovered in the 1960's by Steifel and Scheifele.

2.12 General Canonical Transformations

Not all canonical transformations are easily expressed with generating functions. A more modern approach to canonical transformations begins by defining the phase space state vector as

$$\mathbf{x}^T = \{q_i, \, p_i\} \tag{2.147}$$

Then, Hamilton's equations

$$\dot{q}_i = \frac{\partial H}{\partial p_i}, \quad \dot{p}_i = -\frac{\partial H}{\partial q_i} \tag{2.148}$$

can be written as

$$\dot{\mathbf{x}} = Z\frac{\partial H}{\partial \mathbf{x}} \tag{2.149}$$

The matrix Z which makes the above equation possible is

$$Z = \left\{ \begin{array}{cc} 0 & I \\ -I & 0 \end{array} \right\} \tag{2.150}$$

where each block has order N, the number of degrees of freedom. The matrix Z has several very special properties, the most important of which are

$$-Z = Z^T = Z^{-1} \tag{2.151}$$

All of these are quite simple to verify.

Now, suppose we have a transformation of the phase space from the variables $\mathbf{x}$ to the new variables $\mathbf{y}$. Write the transformation as

$$\mathbf{x} = \mathbf{f}(\mathbf{y}) \tag{2.152}$$

As the transformation above does not involve any time dependence, we will assume that the new Hamiltonian K is

$$K(\mathbf{y}) = H(\mathbf{f}(\mathbf{y})) \tag{2.153}$$

We wish to determine what conditions on the transformation $\mathbf{f}$ are necessary in order that the new coordinates and momenta $\mathbf{y}$ are canonical. That is, they obey Hamilton's equations in the new variables

$$\dot{\mathbf{y}} = Z\frac{\partial K}{\partial \mathbf{y}} \tag{2.154}$$

To determine these conditions, we begin with the purported new form for Hamilton's equations, (2.154), and we will convert them back to the old variables to determine if they are canonical. A derivative of the transformation equations (2.152) gives

$$\dot{\mathbf{x}} = \frac{\partial \mathbf{f}}{\partial \mathbf{y}}\dot{\mathbf{y}} \tag{2.155}$$

Similarly, the gradient of the new Hamiltonian is

$$\frac{\partial K}{\partial \mathbf{y}} = \frac{\partial}{\partial \mathbf{y}}H(\mathbf{f}(\mathbf{y})) = \left(\frac{\partial \mathbf{f}}{\partial \mathbf{y}}\right)^T \frac{\partial H}{\partial \mathbf{x}} \tag{2.156}$$

Substituting these into (2.154) gives

$$\dot{\mathbf{y}} = \left(\frac{\partial \mathbf{f}}{\partial \mathbf{y}}\right)^{-1} \dot{\mathbf{x}} = Z\left(\frac{\partial \mathbf{f}}{\partial \mathbf{y}}\right)^T \frac{\partial H}{\partial \mathbf{x}} \tag{2.157}$$

Solving for $\dot{\mathbf{x}}$ yields

$$\dot{\mathbf{x}} = \frac{\partial \mathbf{f}}{\partial \mathbf{y}}Z\left(\frac{\partial \mathbf{f}}{\partial \mathbf{y}}\right)^T \frac{\partial H}{\partial \mathbf{x}} \tag{2.158}$$

Comparing this to (2.149), the transformation is canonical if

$$\frac{\partial \mathbf{f}}{\partial \mathbf{y}}Z\left(\frac{\partial \mathbf{f}}{\partial \mathbf{y}}\right)^T = Z \tag{2.159}$$

A matrix J which satisfies

$$JZJ^T = Z \tag{2.160}$$

is called *symplectic.* For a transformation of variables $\mathbf{x} = \mathbf{f}(\mathbf{y})$ to be canonical, its Jacobian matrix $J = \partial\mathbf{f}/\partial\mathbf{y}$ must be symplectic. The argument above is reversible, so this is a necessary and sufficient condition.

2.13 References

This chapter especially assumes a working knowledge of basic Lagrangian and Hamiltonian dynamics. There are many good references for these topics, which form the core knowledge of analytical mechanics.

The full details of the Hamilton–Jacobi solution of the two body problem can be found in both Goldstein ("Classical Mechanics", Addison-Wesley Publishing, Reading, Massachusetts, 1959) and Szebehely ("Theory of Orbits", Academic Press, New York, 1967). The actual integrals to obtain the new coordinates require the use of the Cauchy integral theorem on the complex plane.

The discovery that the three dimensional version of the two body problem can be transformed into a *four* dimensional harmonic oscillator is covered in great detail in "Linear and Regular Celestial Mechanics", by Stiefel and Scheifle, Springer-Verlag, New York, 1971.

2.14 Problems

Problem 1. Verify the transformation laws for the F_3 and F_4 generating functions by completing the derivations of section 2.3.

Problem 2. Show that

$$F_2 = \sum q_i P_i$$

is the identity transformation, while the generating function

$$F_1 = \sum q_i Q_i$$

"almost" swaps the role of coordinates and momenta.

Problem 3. The Hamiltonian for a one dimensional harmonic oscillator is

$$H = \frac{1}{2m}p^2 + \frac{k}{2}q^2$$

If we take the generating function

$$F_1 = \frac{1}{2}\sqrt{km}\, q^2 \ \cot Q$$

find expressions for p and P, and invert these to find q and p in terms of the new variables Q and P. Show that the new Hamiltonian is

$$K = \sqrt{\frac{k}{m}} P$$

Solve the system in the new variables.

Problem 4. Taking the Hamiltonian for the one dimensional harmonic oscillator from the previous problem, show that Hamilton's principal function is

$$S = \sqrt{km} \int \left(\frac{2P}{k} - q^2 \right)^{1/2} dq - Pt$$

Show that the transform relation for Q yields the solution

$$q = \sqrt{\frac{2P}{k}} \sin \left(\sqrt{\frac{k}{m}} (t + Q) \right)$$

Calculate the required partial derivatives before performing the integrations. (A cosine will also work above!) Demonstrate that P is the total energy, and show that the new Hamiltonian is identically zero.

Problem 5. Show that an integral of the motion $I(p, q, t)$ for a Hamiltonian system H obeys the partial differential equation

$$[H, I] + \frac{\partial I}{\partial t} = 0$$

where the *Poisson Bracket* is defined as

$$[f, g] = \sum_{k=1}^{N} \left(\frac{\partial f}{\partial p_k} \frac{\partial g}{\partial q_k} - \frac{\partial f}{\partial q_k} \frac{\partial g}{\partial p_k} \right)$$

Problem 6. Show that the dynamics linearization matrix in the equations of variation is

$$A = Z \frac{\partial^2 H}{\partial \mathbf{x}^2}$$

for a canonical system. Show that the propagation law for the state transition matrix Φ:

$$\dot{\Phi} = Z \frac{\partial^2 H}{\partial \mathbf{x}^2} \Phi$$

means that $\Phi(t, t_0)$ is symplectic if $\Phi(t_0, t_0)$ is symplectic. Show that $\Phi(t_0, t_0)$ is always symplectic, and therefore show that $\Phi(t, t_0)$ is always symplectic.

Problem 7. Since we saw above that $\Phi(t, t_0)$ is symplectic, we must have the symplectic condition $\Phi Z \Phi^T = Z$ satisfied. Write this as

$$Z \Phi^T = \Phi^{-1} Z$$

and taking the determinant of both sides, conclude that Φ and Φ^{-1} have the same determinant. Since the determinant of a matrix is the product of its eigenvalues, note that this is consistent with the eigenvalues of a symplectic matrix occurring as inverse pairs: if a is an eigenvalue, so is $1/a$.

Problem 8. Show that the transformation from the Delaunay elements G, g to a new coordinate $Q = \sqrt{2G}\sin g$ and a new momentum $P = \sqrt{2G}\cos g$ is canonical, by computing the Jacobian J of this transformation, and then verifying that $JZJ^T = Z$. Is it possible to give a generating function for this transformation? Find a similar canonical transformation involving the Delaunay momentum H and coordinate h, for the inclination and node.

Problem 9. Let the original coordinate / momentum pair be $\mathbf{x}^T = (g, G)$, and the new variables be

$$\mathbf{y} = \begin{pmatrix} \varsigma \\ \eta \end{pmatrix} = \mathbf{f}(\mathbf{x}) = \begin{pmatrix} \sqrt{2G}\sin g \\ \sqrt{2G}\cos g \end{pmatrix}$$

Use the test of section 2.12 to show that this does represent a canonical transformation. Can this transform be obtained through a generating function?

Chapter 3

Variation of Elements

3.1 Introduction

For several hundred years, celestial mechanics was cursed with very accurate data and (by modern standards) pitifully inadequate computational resources. Even the Royal Observatory at Greenwich, which in the 1800's had several hundred computors (humans), could not support the numerical integration of the equations of motion for an orbit. In this environment, methods had to be developed to generate accurate predictions without requiring very large computational burdens. In particular, numerical integration of equations of motion requires short time steps, so in going from t_1 to t_2 one must calculate the orbit at many intermediate times. This is avoided in general perturbation theory, where an approximate analytic solution to the orbital behavior is obtained in the form of an infinite series expansion.

This method owes its success to the fact that most orbits in the real world are very nearly two body systems. That is, the largest terms in the equations of motion are those arising from two body motion, and terms from other forces are small. If the magnitude of the extra forces are small, one might expect that their effects would also be small. This is not always true, and we will later see examples of the failure of this assumption. However, it is usually a very good assumption. It is the *fundamental assumption of perturbation theory.*

To make use of the fact that the orbit will probably be two – body – like, we need to formulate equations of motion for orbital systems which are adapted to this fact. That is, we will incorporate the assumption of two – body – like motion in the choice of coordinates for the problem. If we use as variables the two body orbital elements, then the two body contributions to the equations of motion vanish: the orbital elements are constant in two body motion. Any terms left in the equations of motion will be due solely to the presence of non – two body forces in the system. It is the formulation of these equations of motion which is the subject of this chapter.

3.2 Variation of Elements

We know how to solve the two body problem exactly in closed form. Most systems encountered in orbital mechanics involve only small changes from two body motion. For example, we could write the equations of motion in the usual rectangular coordinates

$$m\ddot{\mathbf{r}} = -\frac{\mu m \mathbf{r}}{r^3} + \mathbf{F}_p \tag{3.1}$$

where $\mathbf{F}_p$ is the perturbing force, and numerically integrate. However, the two body terms would dominate on the right hand sides of (3.1). We would throw away our knowledge of the solution to the two body problem by doing this. Furthermore, if the effects of the perturbing forces $\mathbf{F}_p$ are small, then we would have to carry many significant figures in the numerical integration of (3.1) to even see the effects of the perturbations. Otherwise, the integration would just produce, at a vastly increased cost, an arc of a Keplerian orbit. It was for this reason that numerical integration was not used for much of the history of celestial mechanics. This is now in the process of changing, since computation is now becoming very cheap, and it is fairly easy to program the numerical integration of (3.1). But the analytical techniques still offer very valuable insight into the character of the solution.

The central idea in the variation of elements is to switch variables and use the two body elements as the new coordinates. This idea has occurred in several fields, and in celestial mechanics used to go under the confusing name of "variation of arbitrary constants". (The "constants" would only be constant in the absence of perturbations, but the possibility of confusion is obvious.) In the theory of differential equations, this technique is called the variation of parameters. Since the six classical orbital elements $(a, e, i, \Omega, \omega, T_o)$ can be exchanged for the physical radius vector and velocity vector $(\mathbf{r}, \mathbf{v})$, the classical elements are one possible replacement for the physical rectangular coordinates used in equation (3.1). Furthermore, we know how to write the equations of motion for the classical elements in the case of pure two body motion:

$$\dot{a} = 0\ ,\ \dot{e} = 0\ ,\ \dot{i} = 0\ ,\ \dot{\Omega} = 0\ ,\ \dot{\omega} = 0\ ,\ \dot{T}_o = 0 \tag{3.2}$$

That is, in the absence of perturbing forces, all six elements are constant. This will no longer be true when there are perturbing forces in the system. However, the dominant two body term in (3.1) will not be present in the analogue (3.2) for perturbed systems, since it will have been absorbed into the definition of the coordinates.

The solution to the element equations will give the six classical elements as functions of time, since with perturbations included they will no longer be constant. However, at any given moment, say t_1, the values of the classical elements $a(t_1), e(t_1), i(t_1)$, etc, describe the two body orbit that the object would follow if all perturbations were to vanish at that instant. That is, the elements describe the two body orbit that the object is instantaneously following at that one moment of time t_1. Given the values of the classical elements at

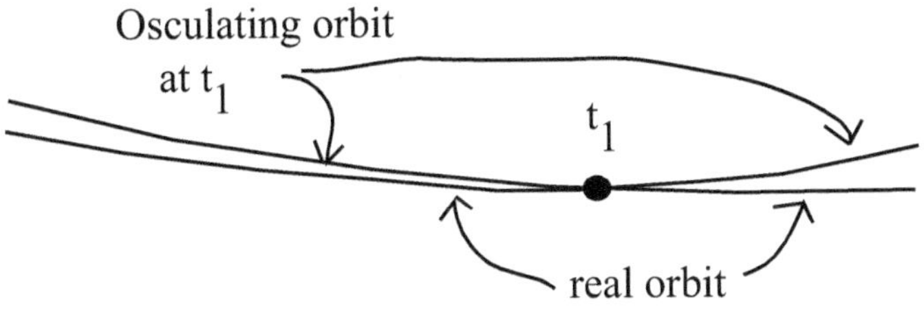

Figure 3.1: Real and Osculating Orbit

that time, we can find the physical position and velocity vectors from the two body relations as if there were no perturbations present in the system at all. The two body problem solution has become a coordinate transformation, from elements to physical variables $(\mathbf{r}, \mathbf{v})$ and back again.

The instantaneous classical elements are called *osculating elements* for the orbit. At any moment, they describe a two body orbit which goes through the same point as the real orbit, so the true and fictitious two body orbits must meet at this point, as shown in Figure 3.1. Furthermore, the instantaneous two body orbit produces the same velocity vector as the real orbit at this time. If only the directions of the velocities are the same, the two body orbit and the real orbit would be tangent at this point. However, the speeds are also the same at this point, so their relationship is more intimate than tangency. The two body orbit and the real orbit are said to 'kiss' (the literal meaning of osculate) at the time t_1. At another time, there will be another osculating orbit, slightly different, which will produce correct predictions for that time, and that time alone.

Thus, the method of producing predictions becomes a two step process. The perturbation equations must be found (and solved) to obtain the osculating orbit as a function of time. Posing these equations is the subject of this chapter. Once the solution is obtained, the values of the osculating elements can be used in the two body orbit equations just as if there were no perturbations at all. The only restriction is that these elements now change slowly with time, and the correct value of t must be used.

3.3 Planetary Equations – Force Form

Suppose a satellite experiences a force $\mathbf{F}_p$ besides the two body force. Any changes in the six classical elements will be due to this force alone, since the two body elements are constant when the only force in the system is the two body gravitational term. If the perturbing force $\mathbf{F}_p$ arises from a potential,

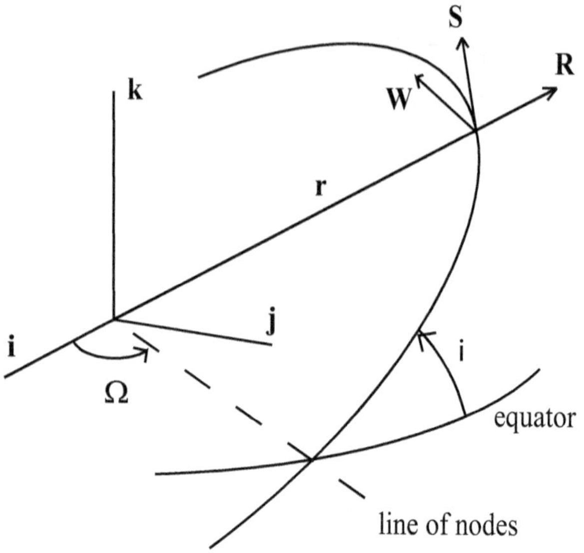

Figure 3.2: Perturbing Force Geometry

it is generally advantageous to use the alternate form of Lagrange's planetary equations, derived in a later section; or better still, a Hamiltonian form of the equations of motion. However, for forces like air drag for which we cannot write a potential, the force component form of Lagrange's planetary equations are still useful. They are also very useful when we ourselves supply the small force, as, for example, in an ion drive spacecraft.

Now, we may not know the mass of the satellite itself. This is not at all uncommon for artificial satellites launched by unfriendly governments, and is the general rule in astronomy when a new asteroid, moon, or comet is discovered. Thus, we will find it useful to work with the components of the perturbing acceleration $\mathbf{a}_p$ rather than the physical force.

Introduce a coordinate frame with a radial unit vector $\mathbf{R}$, a vector normal to the orbit $\mathbf{W}$, and a local horizon vector $\mathbf{S} = \mathbf{W} \times \mathbf{R}$, as shown in Figure 3.2. Note that $\mathbf{S}$ is along the velocity vector only if the orbit is circular. Since these axes differ from the usual $\mathbf{P}, \mathbf{Q}, \mathbf{W}$ orbit frame only by the extra rotation by ν about the orbit normal, we can use the standard rotation matrix to the orbital frame (with the modification that $\omega \to (\omega + \nu)$ to obtain

$$\begin{aligned} \mathbf{R} &= (\cos\Omega\cos(\omega+\nu) - \sin\Omega\sin(\omega+\nu)\cos i)\mathbf{i} \\ &+ (\sin\Omega\cos(\omega+\nu) + \cos\Omega\sin(\omega+\nu)\cos i)\mathbf{j} \end{aligned} \tag{3.3}$$

$$\begin{aligned}
& + \quad (\sin(\omega+\nu)\sin i)\mathbf{k} \\
\mathbf{S} &= (-\cos\Omega\sin(\omega+\nu) - \sin\Omega\cos(\omega+\nu)\cos i)\mathbf{i} && (3.4)\\
& + \quad (-\sin\Omega\sin(\omega+\nu) + \cos\Omega\cos(\omega+\nu)\cos i)\mathbf{j} \\
& + \quad (\cos(\omega+\nu)\sin i)\mathbf{k} \\
\mathbf{W} &= \sin i\sin\Omega\mathbf{i} - \sin i\cos\Omega\mathbf{j} + \cos i\mathbf{k} && (3.5)
\end{aligned}$$

We will resolve the perturbing acceleration along these axes as

$$\mathbf{a}_p = a_r\mathbf{R} + a_s\mathbf{S} + a_w\mathbf{W} \tag{3.6}$$

We wish to find equations of motion for the orbital elements which only show change when perturbing accelerations are present. Each of these elements $\mathcal{E}$ can be written in terms of the instantaneous position and velocity vectors $\mathcal{E}(\mathbf{r},\mathbf{v})$. If we calculated the rate of change of one of these orbital elements, using $\dot{\mathbf{r}} = \mathbf{v}$ and $\dot{\mathbf{v}} = -\mu\mathbf{r}/r^3$,

$$\frac{d\mathcal{E}}{dt} = \frac{\partial\mathcal{E}}{\partial\mathbf{r}}\cdot\mathbf{v} + \frac{\partial\mathcal{E}}{\partial\mathbf{v}}\cdot\left(-\frac{\mu\mathbf{r}}{r^3}\right) = 0 \tag{3.7}$$

we obtain zero, since all six of these elements are constant in the two body problem. When perturbations are present, we still have $\dot{\mathbf{r}} = \mathbf{v}$, but now $\dot{\mathbf{v}} = -\mu\mathbf{r}/r^3 + \mathbf{a}_p$. The time rate of change of an element becomes

$$\begin{aligned}
\frac{d\mathcal{E}}{dt} &= \frac{\partial\mathcal{E}}{\partial\mathbf{r}}\cdot\mathbf{v} + \frac{\partial\mathcal{E}}{\partial\mathbf{v}}\cdot\left(-\frac{\mu\mathbf{r}}{r^3} + \mathbf{a}_p\right) && (3.8)\\
&= \frac{\partial\mathcal{E}}{\partial\mathbf{v}}\cdot\mathbf{a}_p
\end{aligned}$$

since the two body part vanishes. So, to obtain equations of motion for the orbital elements, we need only take their definitions and calculate time derivatives, holding the position vector "constant", and replacing the time derivative of the velocity vector with $\mathbf{a}_p$. The two body contributions from $d\mathbf{v}/dt$ and $d\mathbf{r}/dt$ can be ignored. If instead we substitute $d\mathbf{v}/dt = \mathbf{a}_p$, then we are calculating the rate of change of the orbital element due to the perturbing acceleration alone. The two body contribution adds nothing, since, of course, the result would be zero.

As a beginning, take the energy equation

$$E = \frac{1}{2}\mathbf{v}\cdot\mathbf{v} - \frac{\mu}{r} \tag{3.9}$$

and calculate its rate of change:

$$\frac{dE}{dt} = \mathbf{v}\cdot\mathbf{a}_p \tag{3.10}$$

The perturbing acceleration does not instantaneously change the position vector. The above equation expresses the fact that the rate of change of the energy per unit mass is the rate at which the perturbing acceleration does work on the satellite (per unit mass). Now, the velocity vector of the satellite is

$$\mathbf{v} = \dot{r}\mathbf{R} + r\dot{\nu}\mathbf{S} \tag{3.11}$$

and the semimajor axis is $a = -\mu/2E$, so its rate of change is given by $da/dt = (2a^2/\mu)dE/dt$. These results can be combined to give

$$\frac{da}{dt} = \frac{2a^2}{\mu}(\dot{r}a_r + r\dot{\nu}a_s) \tag{3.12}$$

This is not quite the final form we desire. If we eliminate the quantities $\dot{r}$ and $\dot{\nu}$ using the two body results $\dot{r} = dr/d\nu\ \dot{\nu}$, and

$$\frac{dr}{d\nu} = r\frac{e\sin\nu}{1+e\cos\nu} \tag{3.13}$$

$$\frac{d\nu}{dt} = \frac{na^2}{r^2}\sqrt{1-e^2} \tag{3.14}$$

We obtain

$$\frac{da}{dt} = \frac{2e\sin\nu}{n\sqrt{1-e^2}}a_r + \frac{2a\sqrt{1-e^2}}{nr}a_s \tag{3.15}$$

This is still somewhat of a hybrid form, since the true anomaly and radius are not orbital elements. However, no closed form expressions for ν or r exists in terms of the elements.

Similarly, the angular momentum of the orbit is given by

$$h = \sqrt{\mu a(1-e^2)} = na^2\sqrt{1-e^2} \tag{3.16}$$

Solving for the eccentricity yields

$$e = \sqrt{1 - \frac{h^2}{\mu a}} \tag{3.17}$$

Taking the time derivative gives

$$\frac{de}{dt} = -\frac{h}{2\mu ae}\left(2\frac{dh}{dt} - \frac{h}{a}\frac{da}{dt}\right) \tag{3.18}$$

Now, we already have an expression for da/dt. To calculate the rate of change of the angular momentum, return to its definition and calculate a time derivative:

$$\frac{d\mathbf{h}}{dt} = \frac{d}{dt}(\mathbf{r}\times\mathbf{v}) = \mathbf{r}\times\mathbf{a}_p = ra_s\mathbf{W} - ra_w\mathbf{S} \tag{3.19}$$

Only the perturbing acceleration is included in the above, since the two body force will produce a zero cross product. The rate of change of the magnitude of **h** will simply be the **W** component of the above, since **h** is, by definition, in the **W** direction. This gives $dh/dt = ra_s$. Inserting this into (3.18) we find

$$\frac{de}{dt} = \frac{\sqrt{1-e^2}\sin\nu}{na}a_r + \frac{\sqrt{1-e^2}}{na^2e}\left(\frac{a^2(1-e^2)}{r} - r\right)a_s \tag{3.20}$$

This is the second of the six Lagrange Planetary equations.

Next, the inclination is given by

$$\cos i = \frac{\mathbf{h}\cdot\mathbf{k}}{h} \tag{3.21}$$

and a time derivative produces

$$-\sin i\frac{di}{dt} = \frac{1}{h}\frac{d\mathbf{h}}{dt}\cdot\mathbf{k} - \frac{1}{h^2}\mathbf{h}\cdot\mathbf{k}\frac{dh}{dt} \tag{3.22}$$

Substituting for known quantities, we get the third Lagrange planetary equation

$$\frac{di}{dt} = \frac{r\cos(\omega+\nu)}{na^2\sqrt{1-e^2}}a_w \tag{3.23}$$

since $\mathbf{W}\cdot\mathbf{k} = \cos i$ and $\mathbf{S}\cdot\mathbf{k} = \sin i\cos(\omega+\nu)$.

The next classical element is the node Ω. This is defined as

$$\cos\Omega = \frac{\mathbf{i}\cdot(\mathbf{k}\times\mathbf{h})}{|\mathbf{n}|} = \frac{(\mathbf{i}\times\mathbf{k})\cdot\mathbf{h}}{|\mathbf{n}|} = -\frac{\mathbf{j}\cdot\mathbf{h}}{|\mathbf{n}|} \tag{3.24}$$

where the nodal vector $\mathbf{n} = \mathbf{k}\times\mathbf{h}$, and we have used the ability to exchange the dot and cross in the scalar double product. A time derivative then gives

$$-\sin\Omega\frac{d\Omega}{dt} = -\frac{\mathbf{j}\cdot d\mathbf{h}/dt}{|\mathbf{n}|} + \frac{\mathbf{j}\cdot\mathbf{h}}{|\mathbf{n}|^2}\frac{d|\mathbf{n}|}{dt} \tag{3.25}$$

Now, $|\mathbf{n}| = h\sin i$, so $dn/dt = \dot{h}\sin i + h\cos i di/dt$, and we already know both derivatives. The quantity $d\mathbf{h}/dt$ is already available from (3.19). Performing the dot products with the aid of (3.3)–(3.5), and substituting into (3.25), we have the Lagrange equation for the motion of the node

$$\frac{d\Omega}{dt} = \frac{r\sin(\omega+\nu)}{na^2\sqrt{1-e^2}\sin i}a_w \tag{3.26}$$

Now, let us tackle the argument of perigee ω next. It is defined as

$$\cos\omega = \frac{\mathbf{n}\cdot\mathbf{e}}{ne} \tag{3.27}$$

Taking a time derivative gives us

$$-\sin\omega\frac{d\omega}{dt} = \frac{1}{ne}(\dot{\mathbf{n}}\cdot\mathbf{e} + \mathbf{n}\cdot\dot{\mathbf{e}}) - \frac{\mathbf{n}\cdot\mathbf{e}}{ne}\left(\frac{1}{n}\frac{dn}{dt} + \frac{1}{e}\frac{de}{dt}\right) \tag{3.28}$$

The only quantity we have not already calculated is $d\mathbf{e}/dt$. The eccentricity vector is defined as

$$\mathbf{e} = \frac{1}{\mu}\left[\left(\mathbf{v}\cdot\mathbf{v} - \frac{\mu}{r}\right)\mathbf{r} - (\mathbf{r}\cdot\mathbf{v})\mathbf{v}\right] \tag{3.29}$$

In taking the time derivative of this quantity, we must remember again that $\mathbf{r}$ is instantaneously constant, and that the derivative of $\mathbf{v}$ is $\mathbf{a}_p$, the perturbing acceleration. This gives

$$\frac{d\mathbf{e}}{dt} = \frac{1}{\mu}\left(2(\mathbf{v}\cdot\mathbf{a}_p)\mathbf{r} - (\mathbf{r}\cdot\mathbf{a}_p)\mathbf{v} - (\mathbf{r}\cdot\mathbf{v})\mathbf{a}_p\right) \tag{3.30}$$

in terms of known quantities. Substituting and simplifying, this gives

$$\begin{aligned}\frac{d\omega}{dt} &= -\frac{\sqrt{1-e^2}\cos\nu}{nae}a_r - \frac{r\cot i\sin(\omega+\nu)}{na^2\sqrt{1-e^2}}a_w \\ &+ \frac{\sqrt{1-e^2}}{nae}\left(1+\frac{1}{1+e\cos\nu}\right)\sin\nu\; a_s\end{aligned} \tag{3.31}$$

Finally, rewrite Kepler's equation

$$M = \sqrt{\frac{\mu}{a^3}}t + M_o = E - e\sin E \tag{3.32}$$

as $M_o = -nt + E - e\sin E$, and take a time derivative to obtain

$$\frac{dM_o}{dt} = \frac{3}{2}\sqrt{\frac{\mu}{a^5}}t\frac{da}{dt} + (1-\cos E)\frac{dE}{dt} \tag{3.33}$$

To obtain the derivative of the eccentric anomaly, turn to the two body relation

$$\cos E = \frac{e+\cos\nu}{1+e\cos\nu} \tag{3.34}$$

and take a derivative to find

$$-\sin E\frac{dE}{dt} = \frac{\dot{e} - \sin\nu\dot{\nu}}{1+e\cos\nu} - \frac{(e+\cos\nu)(\dot{e}\cos\nu - \sin\nu\dot{\nu})}{(1+e\cos\nu)^2} \tag{3.35}$$

This introduces another unknown quantity, $d\nu/dt$. To obtain this quantity, turn to the definition of ν

$$\cos\nu = \frac{\mathbf{e}\cdot\mathbf{r}}{er} \tag{3.36}$$

Again, we must take a funny derivative, holding $\mathbf{r}$ instantaneously constant, and only considering the contribution due to perturbations. This gives

$$-\sin\nu\frac{d\nu}{dt} = \frac{\dot{\mathbf{e}}\cdot\mathbf{r}}{er} - \frac{\mathbf{e}\cdot\mathbf{r}}{e^2r}\frac{de}{dt} \tag{3.37}$$

and all quantities are known or directly calculable. Combination of (3.33) – (3.37) and simplification gives the final Lagrange Planetary equation

$$\begin{aligned}\frac{dM_o}{dt} &= -\frac{1}{na}\left(\frac{2r}{a}-\frac{1-e^2}{e}\cos\nu\right)a_r+\frac{3}{2}t\sqrt{\frac{\mu}{a^5}}\frac{da}{dt}\\ &- \frac{1-e^2}{nae}\left(1+\frac{r}{a(1-e^2)}\right)\sin\nu\, a_s \end{aligned} \tag{3.38}$$

For reference, the complete set is given below:

The Lagrange Planetary Equations Acceleration Component Form

$$\frac{da}{dt}=\frac{2e\sin\nu}{n\sqrt{1-e^2}}a_r+\frac{2a\sqrt{1-e^2}}{nr}a_s \tag{3.39}$$

$$\frac{de}{dt}=\frac{\sqrt{1-e^2}\sin\nu}{na}a_r+\frac{\sqrt{1-e^2}}{na^2e}\left(\frac{a^2(1-e^2)}{r}-r\right)a_s \tag{3.40}$$

$$\frac{di}{dt}=\frac{r\cos(\omega+\nu)}{na^2\sqrt{1-e^2}}a_w \tag{3.41}$$

$$\frac{d\Omega}{dt}=\frac{r\sin(\omega+\nu)}{na^2\sqrt{1-e^2}\sin i}a_w \tag{3.42}$$

$$\begin{aligned}\frac{d\omega}{dt} &= -\frac{\sqrt{1-e^2}\cos\nu}{nae}a_r-\frac{r\cot i\sin(\omega+\nu)}{na^2\sqrt{1-e^2}}a_w\\ &+ \frac{\sqrt{1-e^2}}{nae}\left(1+\frac{1}{1+e\cos\nu}\right)\sin\nu\, a_s\end{aligned} \tag{3.43}$$

$$\begin{aligned}\frac{dM_o}{dt} &= -\frac{1}{na}\left(\frac{2r}{a}-\frac{1-e^2}{e}\cos\nu\right)a_r+\frac{3}{2}t\sqrt{\frac{\mu}{a^5}}\frac{da}{dt}\\ &- \frac{1-e^2}{nae}\left(1+\frac{r}{a(1-e^2)}\right)\sin\nu\, a_s \end{aligned} \tag{3.44}$$

Notice that certain of these equations contain the eccentricity e or the sine of the inclination $\sin i$ as divisors. This is much more than a minor irritant when either of these quantities approach zero. This is another manifestation of the problem with defining the argument of perigee or the right ascension of the node when the eccentricity or the inclination become zero, respectively. In a numerical integration of Lagrange's planetary equations, this could lead the computer to divide by zero. In an analytic effort, this leads to inconvenient infinities in the perturbation equations. Besides, perturbations are supposed to remain small! Transforming the equations for e, ω, i, and Ω to any of the alternative sets mentioned in the first chapter will eliminate these difficulties, or at least move them to a place where they cannot cause trouble.

3.4 Decay Lifetime

The most important use of the Lagrange planetary equations is in deriving approximate solutions to problems not otherwise easily treated. As an example, consider the lifetime of an earth satellite whose orbit is perturbed by atmospheric drag. A satellite in a low orbit suffers a significant acceleration due to drag, given by

$$a_d = \frac{1}{2}\frac{C_D A}{m}\rho v^2 \tag{3.45}$$

where C_D is the ballistic coefficient, A is the presented area of the satellite, and m is its mass. These three quantities are not normally determinable separately, so they are grouped into the single quantity:

$$B^* = \frac{C_D A}{m} \tag{3.46}$$

The other quantities in (3.45) are ρ, the atmospheric density, and v is the satellite velocity.

The first effect of air drag on a satellite orbit is to circularize the orbit. This occurs since the effect of drag is much more pronounced near perigee, when the satellite is closest to the earth. The satellite is slowed at this point, lowering the apogee height until it nearly equals the perigee height. We will assume that the orbit has circularized, and study the behavior of the semimajor axis. With a circular orbit, all the air drag force is in the tangential direction. For simplicity, we will assume that the atmosphere is exponential, with air density given by:

$$\rho = \rho_o \exp\left(-(r - R_\oplus)/h\right) \tag{3.47}$$

where ρ_o is a fictitious base density of the atmosphere, $R_\oplus$ is the radius of the earth, and h is called the atmospheric scale height. At the top of the atmosphere, h is approximately 6 to 8 kilometers. So, assuming a circular orbit, where $a = r$ and the satellite velocity is given by

$$v = \sqrt{\frac{\mu}{r}} \tag{3.48}$$

the equation of motion for the semimajor axis becomes

$$\frac{da}{dt} = -\sqrt{\mu a} B^* \rho_o \exp\left(-(a - R_\oplus)/h\right) \tag{3.49}$$

since the air drag accelerates the satellite in the $-\mathbf{S}$ direction. Notice what we have done to produce the above: the expressions for the speed from a two body circular orbit have been inserted into the right hand side. This is the small perturbation assumption at work, since the error made is proportional to the product of two small quantities: the small air drag force times the small error made in assuming the orbit is circular, when it is really going to be a very slow spiral.

The variables in this equation can be separated to yield the two definite integrals

$$\int_{a_o}^{a} \frac{\exp((a - R_\oplus)/h)}{\sqrt{a}} \, da = -\sqrt{\mu} B^* \rho_o \int_{t_o}^{t} dt \tag{3.50}$$

In perturbation theory, one normally does whatever is necessary to obtain an approximate solution, without, of course, totally gutting the problem of physical significance. Let us begin by changing variables on the left side of (3.50) to altitude above the surface, $H = a - R_\oplus$, to obtain

$$\int_{H_o}^{H} \frac{\exp(H/h)}{\sqrt{R_\oplus + H}} \, dH = -\sqrt{\mu} B^* \rho_o (t - t_o) \tag{3.51}$$

The radical in the denominator is proportional to the satellite velocity, which does not greatly increase during the last hundred kilometers of decay. We can thus approximate this quantity by the root of $R_\oplus$ alone. This avoids expanding the exponential function, which is not desirable since the air density *does* vary enormously during a decay trajectory.

With this simplification, the integral above can be performed by simple processes to yield

$$\frac{h}{\sqrt{R_\oplus}} (\exp(H/h) - \exp(H_o/h)) = -\sqrt{\mu} B^* \rho_o (t - t_o) \tag{3.52}$$

This may be solved to give the behavior of altitude with time

$$H(t) = h \ln \left(\exp(H_o/h) - \frac{\sqrt{\mu R_\oplus} B^* \rho_o}{h} (t - t_o) \right) \tag{3.53}$$

A typical satellite in a low earth orbit might have an initial altitude of from 20 to 40 scale heights. In this case, the first term in the logarithm is huge, and dominates for a very long time. Thus equation (3.53) reduces approximately to the statement $H = H_o$. However, eventually the linear term will become comparable to the first term, and then the altitude decreases very quickly indeed. An estimate of the time to elapse to decay can be obtained by setting the altitude in (3.53) to zero. Since the terminal descent is very rapid, this introduces little error. The satellite reaches zero altitude when the argument of the logarithm function equals unity. Interpreting the time difference $t - t_o$ as the time to elapse to decay, t_d, we easily find

$$t_d = \frac{h}{\sqrt{\mu R_\oplus} B^* \rho_o} (\exp(H_o/h) - 1) \tag{3.54}$$

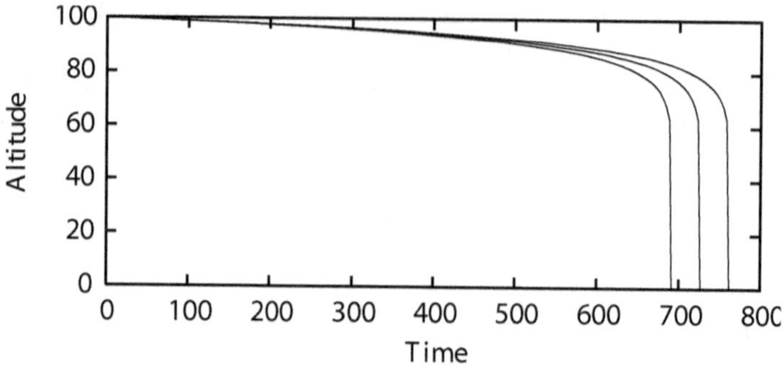

Figure 3.3: Altitude versus Time during final decay

It is in accord with expectations that the satellite lifetime be inversely proportional to both B^* and ρ_o, only one of which is under the control of the engineer. However, the satellite lifetime is an exponentially growing function of initial orbital altitude. So, rather than try to design streamlined satellites, the easiest way to extend the lifetime of a low earth satellite is to place it in a slightly higher orbit. An increase of H_o by one scale height (6 to 8 kilometers) will extend the satellite lifetime by almost a factor of three.

In the analysis above, the most questionable assumption was the model for the air density. At any given moment, the air density is approximately exponential, but the base density of the atmosphere ρ_o and the scale height h may change drastically in a short period of time. This can occur during a solar storm. Also, they change with the phase of the 11 year sunspot cycle. This makes long term decay predictions very inaccurate.

Figure 3.3 shows three decay trajectories as a function of time. They differ only in the scale height h for the atmosphere, and these differed by only 5 percent. At the beginning, it is extremely difficult to tell these trajectories apart. So, even if h were a dependable constant of the atmosphere, it would still be very difficult to predict the final plunge into the atmosphere. This fact renders even an accurate orbital fit in the pre–decay environment of very doubtful value in accurately predicting the final decay time. In practice, it is seldom possible to predict the actual decay date until about a week before decay, and the final decay orbit cannot be predicted until a few hours in advance.

3.5 The Cannonball Problem

As a classical example of setting a problem up for a perturbation theory treatment, let us consider the problem of a cannonball again, only this time including air drag. The dissipative air drag force does not lend itself to canonical treatment, so we will take a classical mechanics approach. We begin with the equations of motion for a projectile including air drag in the vertical / down–range plane:

$$\ddot{\mathbf{r}} = -g\mathbf{k} - \frac{1}{2}B^*\rho v\mathbf{v} \tag{3.55}$$

where $\mathbf{r}^T = (x, \; z)$ locates the projectile. Now, the "solvable" part of this can obviously be taken to be the part without air drag. In this case we can write down the solution by elementary means as

$$\begin{aligned} x(t) &= x_o + v_{xo}t \\ z(t) &= z_o + v_{zo}t - \frac{1}{2}gt^2 \\ \dot{x}(t) &= v_{xo} \\ \dot{z}(t) &= v_{zo} - gt \end{aligned} \tag{3.56}$$

where the subscripted quantities x_o, z_o, v_{xo}, and v_{zo} are the initial conditions at $t = 0$.

Next, we must find a set of *elements* for this problem. Now, if we think of the initial conditions as only initial conditions, we still don't have elements for this problem. On the other hand, x_o, z_o, v_{xo}, and v_{zo} are all constant when air drag is zero. If we define an element vector as $\mathbf{E}^T = (x_o, \; z_o, \; v_{xo}, \; v_{zo})$, and the state vector $\mathbf{X}^T(t) = (x(t), \; z(t), \; \dot{x}(t), \; \dot{z}(t))$, then equations (3.56) give the relation between the elements and the state. The inverse coordinate transform is then easily found to be

$$\begin{aligned} x_o &= x(t) - \dot{x}(t)t \\ z_o &= z(t) - \dot{z}(t)t - \frac{1}{2}gt^2 \\ v_{xo} &= \dot{x}(t) \\ v_{zo} &= \dot{z}(t) + gt \end{aligned} \tag{3.57}$$

which has only state vector elements on the right side. Equations (3.56) and (3.57) are then inverse *coordinate transforms* of each other. They describe the true trajectory in terms of an "osculating parabola".

Now, taking time derivatives of the elements, we have

$$\begin{aligned} \dot{x}_o &= \dot{x}(t) - \dot{x}(t) - \ddot{x}(t)t = -\ddot{x}(t)t \\ \dot{z}_o &= \dot{z}(t) - \dot{z}(t) - \ddot{z}(t)t - gt = -\ddot{z}(t)t - gt \\ \dot{v}_{xo} &= \ddot{x}(t) \\ \dot{v}_{zo} &= \ddot{z}(t) + g \end{aligned} \tag{3.58}$$

If we now insert $\ddot{x}(t) = 0$, $\ddot{z}(t) = -g$, we obtain zeros on the right of all four of equations (3.58). In other words, we have confirmed that $\mathbf{E}$ really is a set of elements for the cannonball problem without air drag.

On the other hand, we can substitute the acceleration of gravity plus air drag from equation (3.55). Now, with air drag included, we have

$$\ddot{x}(t) = -\frac{1}{2}B^*\rho\sqrt{\dot{x}^2 + \dot{z}^2}\,\dot{x} \tag{3.59}$$

and

$$\ddot{z}(t) = -g - \frac{1}{2}B^*\rho\sqrt{\dot{x}^2 + \dot{z}^2}\,\dot{z} \tag{3.60}$$

Inserting these into equations (3.58), the "parabola" part of the equations of motion cancel as before, and we have

$$\dot{x}_o = +\frac{1}{2}B^*\rho\sqrt{\dot{x}^2 + \dot{z}^2}\,\dot{x}t \tag{3.61}$$

$$\dot{z}_o = +\frac{1}{2}B^*\rho\sqrt{\dot{x}^2 + \dot{z}^2}\,\dot{z}t \tag{3.62}$$

$$\dot{v}_{xo} = -\frac{1}{2}B^*\rho\sqrt{\dot{x}^2 + \dot{z}^2}\,\dot{x} \tag{3.63}$$

$$\dot{v}_{zo} = -\frac{1}{2}B^*\rho\sqrt{\dot{x}^2 + \dot{z}^2}\,\dot{z} \tag{3.64}$$

These are perturbation equations for the rate of change of the initial conditions under the influence of air drag. The right sides are all zero, and the elements all constant, should air drag be negligible.

On the other hand, equations (3.61) – (3.64) are not yet proper differential equations. The left sides are rates of change of the elements, while the right sides still involve state vector components. The right sides *should* involve the *elements* themselves. To insert the elements, remember that equations (3.56) are now just a coordinate transformation between the state vector components and the "osculating parabola" elements. Then

$$v = \sqrt{\dot{x}^2 + \dot{z}^2} = \left(v_{xo}^2 + v_{zo}^2 - 2v_{zo}gt + g^2t^2\right)^{1/2} \tag{3.65}$$

and with this abbreviation the element equations become:

$$\dot{x}_o = \frac{1}{2}B^*\rho v v_{xo} t \tag{3.66}$$

$$\dot{z}_o = \frac{1}{2}B^*\rho v\left(v_{zo} - gt\right)t \tag{3.67}$$

$$\dot{v}_{xo} = -\frac{1}{2}B^*\rho v v_{xo} \tag{3.68}$$

$$\dot{v}_{zo} = -\frac{1}{2}B^*\rho v\left(v_{zo} - gt\right) \tag{3.69}$$

These are proper differential equations, and are highly non–linear also. It is highly doubtful that they can be solved in closed form.

3.6 Variation of Canonical Elements

One excellent reason for working with canonical variables is that the perturbation equations are trivial to derive. Suppose we have a problem in which the two body part of the Hamiltonian dominates. If there are perturbing forces, the full system Hamiltonian takes the form

$$H = H_{2b} + H_p \tag{3.70}$$

In general, the extra term $H_p = V_p$ is the potential energy of the perturbing forces in the system. There are no extra kinetic energy terms, since these are already included in the two body part.

Now, suppose the system (3.70) is transformed into the Delaunay elements. These are a set of canonical variables which essentially solve the two body problem. In these variables the system (3.70) takes the form

$$K = -\frac{\mu^2}{2L^2} + K_p(L, G, H, l, g, h, t) \tag{3.71}$$

where K_p is just H_P transformed to the new variables. These immediately lead to the equations of motion

$$\begin{aligned} \frac{dL}{dt} &= -\frac{\partial K_p}{\partial l} + Q_L, & \frac{dl}{dt} &= \frac{\mu^2}{L^3} + \frac{\partial K_p}{\partial L} + Q_l \\ \frac{dG}{dt} &= -\frac{\partial K_p}{\partial g} + Q_G, & \frac{dg}{dt} &= \frac{\partial K_p}{\partial G} + Q_g \qquad (3.72) \\ \frac{dH}{dt} &= -\frac{\partial K_p}{\partial h} + Q_H, & \frac{dh}{dt} &= \frac{\partial K_p}{\partial H} + Q_h \qquad (3.73) \end{aligned}$$

These are, of course, just Hamilton's equations for the system (3.71), with canonical forces Q_i added. We only need these forces if there are non–potential forces acting in the system, and we would need forces on both the coordinate and momenta equations since canonical transformations have occurred.

These are a set of perturbation equations, just as useful as the Lagrange planetary equations but much simpler in form. They are also the same for other dynamical systems, so long as the dominant portion of the Hamiltonian is modified as appropriate.

3.7 Canonical Perturbations - The Pendulum

Canonical transformations can be used to produce perturbation equations for systems other than the two body problem. Consider the pendulum shown in Figure 3.4, with mass m and length l. It has a Lagrangian per unit mass given by

$$L = \frac{1}{2}l^2\dot{\theta}^2 + gl\cos\theta \tag{3.74}$$

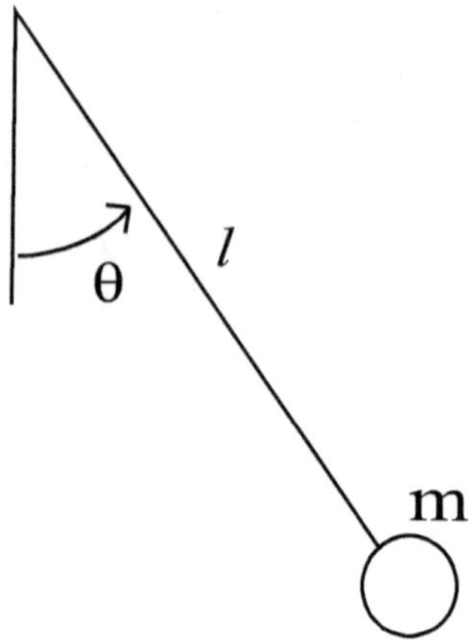

Figure 3.4: The Simple Pendulum

So the generalized momentum is $p = l^2\dot{\theta}$, and the system Hamiltonian becomes

$$H = \frac{1}{2l^2}p^2 - gl\cos\theta \tag{3.75}$$

If we substitute the expansion for the cosine function, the Hamiltonian can be written in the convergent Taylor's series

$$H = \frac{1}{2l^2}p^2 + gl\frac{\theta^2}{2} - gl\sum_{n=2}^{\infty}\frac{(-1)^n}{(2n)!}\theta^{2n} \tag{3.76}$$

The constant term leading the cosine expansion can be dropped in the Hamiltonian, since it will not effect the equations of motion. The first two terms in the above are the harmonic oscillator approximation for the pendulum. It is common in introductory dynamics courses to assume small amplitude oscillations and "solve" the pendulum problem. We can use this limiting case as the basis for a perturbation theory attack on the entire problem. The series expansion above is the necessary first step to isolate the "solvable" part from the "perturbation" part.

Let us apply the generating function

$$F_1 = \frac{1}{2}\sqrt{gl^3}\,\theta^2\,\cot Q \tag{3.77}$$

similar to the generating function applied in a problem in the previous chapter. The transform laws give

$$p = \frac{\partial F_1}{\partial \theta} = \sqrt{gl^3}\,\theta\,\cot Q \tag{3.78}$$

$$P = -\frac{\partial F_1}{\partial Q} = \frac{1}{2}\sqrt{gl^3}\;\theta^2\,\csc^2 Q \tag{3.79}$$

The second of these can be solved to give the coordinate θ in terms of the new coordinate and momentum

$$\theta = \sqrt{\frac{2P}{\sqrt{gl^3}}}\,\sin Q \tag{3.80}$$

The coefficient in front of the sine function is the amplitude of the oscillation, which we are assuming to be small. This implies that the new momentum P is also small.

Continuing with the transformation, the new Hamiltonian is found to be

$$K = \sqrt{\frac{g}{l}}P - gl\sum_{n=2}^{\infty}\frac{(-1)^n}{(2n)!}\left(\frac{2P}{\sqrt{gl^3}}\right)^n \sin^{2n} Q \tag{3.81}$$

The first term of this expression is all that remains of the "harmonic oscillator" part of the pendulum Hamiltonian. The extra terms represent the perturbation to the harmonic oscillator when the amplitude is not small.

Now, if we simply restricted ourselves to the harmonic oscillator part, the system would have Hamiltonian

$$K = \sqrt{\frac{g}{l}}P \tag{3.82}$$

which would give equations of motion easily solved to yield

$$P = P_o \tag{3.83}$$

a constant, and

$$Q = \sqrt{\frac{g}{l}}t + Q_o \tag{3.84}$$

a linear function of time. In the presence of the perturbation terms, however, this is no longer true. However, the Hamiltonian (3.81) gives equations of motion for the elements P and Q for an osculating harmonic oscillator solution to the pendulum problem. The first order solution to this problem will be treated in a later chapter.

3.8 Planetary Equations: The Potential Form

Prior to the acceptance of the canonical form of perturbation equations during this century by the celestial mechanics community, the standard form of perturbation equations used the classical elements $(a, e, i, \Omega, \omega, M_o)$ for the two body problem, but employing partials of a potential function instead of actual physical acceleration components. Most perturbative effects arise from potential forces, and series expansions are required to express the potential (or the force) in terms of the classical elements. It is thus simpler to expand the potential function once, and take partial derivatives of the expansion than to derive the acceleration components and then be forced to expand each of the components separately. However, Lagrange derived this form of the equations before the scientific community settled on the currently accepted sign conventions for potential energy. So, instead of working with the potential energy for the disturbing force, K_p, the Lagrange Planetary Equations are historically posed in terms of a *disturbing function* R

$$R = -K_p = -V_p \tag{3.85}$$

which is simply the negative of the potential function (per unit mass) for the disturbing forces.

Textbooks on celestial mechanics often derive these equations either through complicated arguments based on the principle of virtual work (see Danby), or through complicated arguments based on Poisson and Lagrange brackets (see Brouwer and Clemence). However, the simplest derivation is to simply change variables in the canonical form of an earlier section.

Begin with the first Delaunay momentum $L = \sqrt{\mu a}$. The canonical equation of motion is $dL/dt = -\partial K_p/\partial l$. But the Delaunay coordinate $l = M = nt + M_o$, so the partial derivative with respect to l is simply replaced with a partial derivative with respect to M_o. Also, taking the derivative of the definition of L, we have $dL/dt = \sqrt{\mu/a}\ \dot{a}/2$. Combining these results, the first Lagrange Planetary equation is

$$\frac{da}{dt} = \frac{2}{na}\frac{\partial R}{\partial M_o} \tag{3.86}$$

where n is the two body mean motion.

Now, the second Delaunay momentum is $G = L\sqrt{1-e^2}$. In the canonical equation of motion $dG/dt = -\partial K_p/\partial g$, the coordinate $g = \omega$, the argument of perigee. Calculating the derivative of the definition of G, we have

$$\dot{G} = \dot{L}\sqrt{1-e^2} - \sqrt{\mu a}(1-e^2)^{-1/2} e\dot{e} \tag{3.87}$$

This is easily solved for $\dot{e}$, yielding

$$\dot{e} = \frac{1-e^2}{\sqrt{\mu a} e}\dot{L} - \frac{\sqrt{1-e^2}}{\sqrt{\mu a} e}\dot{G} \tag{3.88}$$

Replacing the quantities $\dot{L}$ and $\dot{G}$ from Hamilton's equations, we obtain the second Lagrange Planetary equation

$$\frac{de}{dt} = \frac{1-e^2}{na^2 e}\frac{\partial R}{\partial M_o} - \frac{\sqrt{1-e^2}}{na^2 e}\frac{\partial R}{\partial \omega} \tag{3.89}$$

The third momentum is $H = G\cos i$. Calculating a time derivative, we find

$$\dot{H} = \dot{G}\cos i - G\sin i\frac{di}{dt} \tag{3.90}$$

Solving for the rate of change of the inclination, this gives

$$\frac{di}{dt} = \frac{1}{G\sin i}\left(\dot{G}\cos i - \dot{H}\right) \tag{3.91}$$

Substituting the definition of $G = \sqrt{\mu a(1-e^2)}$, and the Hamiltonian relations

$$\dot{G} = -\frac{\partial K_p}{\partial g} = \frac{\partial R}{\partial \omega}, \quad \dot{H} = -\frac{\partial K_p}{\partial h} = \frac{\partial R}{\partial \Omega} \tag{3.92}$$

we find the Lagrange equation for the inclination

$$\frac{di}{dt} = \frac{\cot i}{na^2\sqrt{1-e^2}}\frac{\partial R}{\partial \omega} - \frac{1}{na^2\sqrt{1-e^2}\sin i}\frac{\partial R}{\partial \Omega} \tag{3.93}$$

The remaining three Lagrange equations are found directly from Hamilton's equations for l, g, and h. Now, however, it is necessary to transform the partial derivatives from derivatives with respect to L, G, and H to derivatives with respect to the classical elements. In performing this transformation, it is necessary to remember that L, G, and H are considered independent variables in Hamilton's equations. We must preserve this nicety when transforming the partial derivatives of R.

Begin with the mean anomaly $l = M = nt + M_o$, and let us find an equation for the additive constant M_o. Since the nt part is generated by the two body portion of the Hamiltonian, any changes to M_o must be due to the perturbations. So, Hamilton's equation for l becomes

$$\frac{dM_o}{dt} = \frac{\partial K_p}{\partial L} = -\frac{\partial R}{\partial L} \tag{3.94}$$

and we must transform the partial derivative with respect to L (with G, H held constant) into terms involving derivatives with respect to the elements. Beginning with $L = \sqrt{\mu a}$, we have

$$\partial L = \frac{1}{2}\sqrt{\mu}a^{-1/2}\partial a = \frac{na}{2}\partial a \tag{3.95}$$

Also, we could write $L = G/\sqrt{1-e^2}$, so

$$\partial L = G(1-e^2)^{-3/2}e\partial e = \frac{na^2 e}{1-e^2}\partial e \tag{3.96}$$

However, there is no inclination term, since it is not possible to write L in a form containing i and G and H. So, equation (3.94) becomes

$$\frac{dM_o}{dt} = -\frac{2}{na}\frac{\partial R}{\partial a} - \frac{1-e^2}{na^2 e}\frac{\partial R}{\partial e} \tag{3.97}$$

The second Delaunay coordinate is $g = \omega$. So, the Hamilton's equation is

$$\frac{d\omega}{dt} = \frac{\partial K_p}{\partial G} = -\frac{\partial R}{\partial G} \tag{3.98}$$

Again, G can be written two ways in terms of a single classical element and the independent momenta. The first is $G = L\sqrt{1-e^2}$, which leads to

$$\partial G = -\frac{na^2 e}{\sqrt{1-e^2}}\partial e \tag{3.99}$$

Also, we can write $G = H/\cos i$, which leads to

$$\partial G = na^2\sqrt{1-e^2}\tan i\partial i \tag{3.100}$$

So, again we obtain two terms when transforming the partial derivative in (3.98):

$$\frac{d\omega}{dt} = \frac{\sqrt{1-e^2}}{na^2 e}\frac{\partial R}{\partial e} - \frac{\cot i}{na^2\sqrt{1-e^2}}\frac{\partial R}{\partial i} \tag{3.101}$$

The last variable is the Delaunay coordinate $h = \Omega$, the right ascension of the ascending node. The canonical equation of motion is $dh/dt = \partial K_p/\partial H$, which involves a partial derivative of K_p with respect to H, the conjugate momentum. Now, the definition of this momentum is $H = G\cos i$, and the canonical variables H and G are independent. This implies that $\partial H = -G\sin i\,\partial i$, or

$$-\frac{\partial R}{\partial H} = \frac{1}{G\sin i}\frac{\partial R}{\partial i} \tag{3.102}$$

Thus, we arrive at the Lagrange Planetary equation for the motion of the node:

$$\frac{d\Omega}{dt} = \frac{1}{na^2\sqrt{1-e^2}\sin i}\frac{\partial R}{\partial i} \tag{3.103}$$

So, the complete set of Lagrange Planetary Equations, in their potential (disturbing function) form, is:

The Lagrange Planetary Equations
Disturbing Function Form

$$\frac{da}{dt} = \frac{2}{na}\frac{\partial R}{\partial M_o} \tag{3.104}$$

$$\frac{de}{dt} = \frac{1-e^2}{na^2 e}\frac{\partial R}{\partial M_o} - \frac{\sqrt{1-e^2}}{na^2 e}\frac{\partial R}{\partial \omega} \tag{3.105}$$

$$\frac{di}{dt} = \frac{\cot i}{na^2\sqrt{1-e^2}}\frac{\partial R}{\partial \omega} - \frac{1}{na^2\sqrt{1-e^2}\sin i}\frac{\partial R}{\partial \Omega} \tag{3.106}$$

$$\frac{d\Omega}{dt} = \frac{1}{na^2\sqrt{1-e^2}\sin i}\frac{\partial R}{\partial i} \tag{3.107}$$

$$\frac{d\omega}{dt} = \frac{\sqrt{1-e^2}}{na^2 e}\frac{\partial R}{\partial e} - \frac{\cot i}{na^2\sqrt{1-e^2}}\frac{\partial R}{\partial i} \tag{3.108}$$

$$\frac{dM_o}{dt} = -\frac{2}{na}\frac{\partial R}{\partial a} - \frac{1-e^2}{na^2 e}\frac{\partial R}{\partial e} \tag{3.109}$$

Note that these equations also contain the eccentricity e and the sine of the inclination i as divisors. These are the usual problems with the definitions of the argument of perigee when $e = 0$, and the definition of the node when $i = 0$. The alternate variable sets discussed in the previous chapters eliminate these difficulties.

3.9 References

The subject of variation of constants, or perturbation equations of motion, has a very long history. As extra reading, consult chapters 11 of either Brouwer and Clemence, *Methods of Celestial Mechanics,* Academic Press, 1961, or Danby, "Fundamentals of Celestial Mechanics", MacMillan, New York, 1962, and chapter 9 Bate, Mueller, and White, "Fundamentals of Astrodynamics", Dover, New York, 1971.

3.10 Problems

Problem 1. Transform the potential form of Lagrange's Planetary Equations to find expressions for the elements

$$\frac{dh}{dt}, \frac{dk}{dt}, \frac{d\chi}{dt}, \frac{d\psi}{dt}$$

Remember to transform the partial derivatives as well, since the disturbing function is now assumed to be $R(a, M_o, h, k, \chi, \psi)$. Show that the difficulties with small eccentricities and inclinations disappear, with new troubles for $i \rightarrow 180^o$.

Problem 2. A harmonic oscillator with a "hard spring" has a Hamiltonian

$$H = \frac{1}{2m}p^2 + \frac{k}{2}q^2 + \epsilon q^4$$

where the perturbation coefficient $\epsilon << k$. Apply the Hamilton – Jacobi transformation for the harmonic oscillator (see problems, Chapter 2) to the entire Hamiltonian above, and find the new Hamiltonian and equations of motion for the harmonic oscillator "elements".

Problem 3. An ion drive spacecraft is in a virtually circular orbit, located by the true argument of latitude $u = \omega + \nu$. It has (low) acceleration A, and is trying to change both its inclination i and semimajor axis a. It executes a pitching motion, so that in the orbital reference frame the acceleration components are

$$a_r = 0, \quad a_s = A\cos\theta, \quad a_p = A\sin\theta$$

The variable θ is called a control variable, since we do not have dynamics for it, but it is under our direct control

Set up the Lagrange Planetary equations for the semimajor axis, inclination, eccentricity, and node. If the control profile is given by

$$\theta = u$$

calculate the changes in these four variables over one orbit. If the pitch control pattern is shifted to $\theta = u + \pi/2$, exchanging the sine and cosine functions, what change in the elements now occurs?

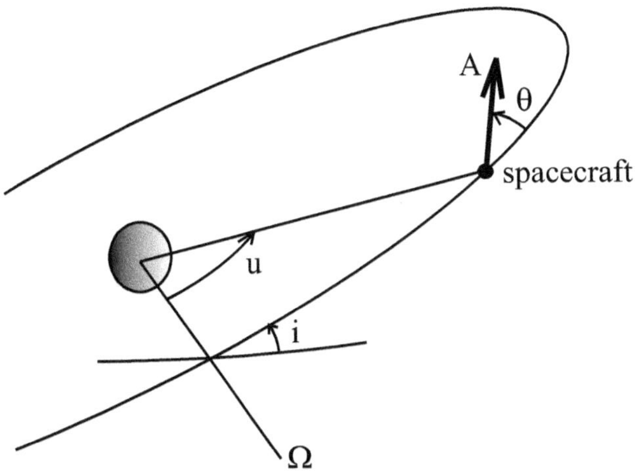

Figure 3.5: Ion drive spacecraft in a nearly circular orbit.

Chapter 4

Perturbing Forces

4.1 Introduction

The solvable problem in celestial mechanics is the two body problem, involving two masses interacting through Newtonian point mass gravity. Besides this force, there are several perturbing forces which commonly occur in orbital mechanics problems. Historically, the force most usually treated has been the interaction between an orbiting body and a "third" gravitating mass. This is the N body point mass problem, which occurs in the solar system. Objects in the solar system are generally so widely separated that their gravitational fields are essentially point mass fields.

However, with the advent of artificial earth satellites, several other perturbing forces have assumed new importance. The gravitational field of the earth is not that of a perfect point mass. Because of its' rotation, the earth bulges at the equator. Also, the continental blocks, ocean basins, mountain ranges, etc produce local deviations from a spherical Newtonian point mass field. While the existence of these effects has been known for a long time, these deviations are unimportant at large distances from the earth. Only with artificial satellites has it become necessary to include and possible to detect these effects.

Another perturbing force which has assumed new importance in the space age is air drag. No natural objects in the solar system orbit their primary bodies close enough to make air drag an important factor in their orbital dynamics. Also, natural objects are usually many kilometers in diameter, and by human standards possess enormous masses. This makes them essentially immune to atmospheric drag. However, small low mass earth satellites orbit very close to the atmosphere, and air drag is usually the determining factor in the lifetime of a low earth satellite. The international space station has an enormous area for its total mass, and air drag is a very important factor in the orbit of this vehicle.

Finally, the space environment offers chances to observe forces which cannot be detected in earthly dynamics. The large (100 foot diameter) Echo and Pageos

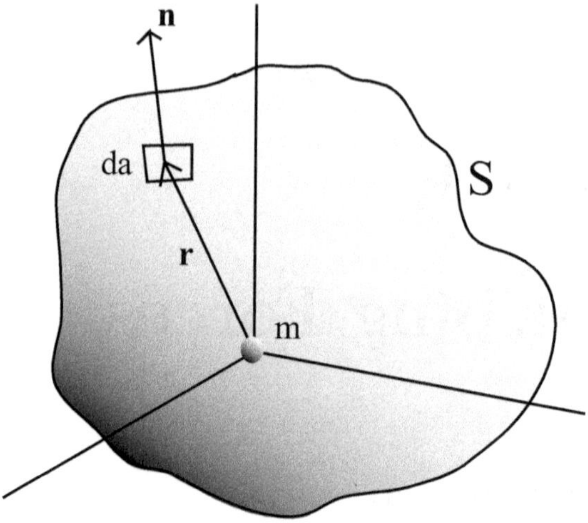

Figure 4.1: Total Field Calculation

balloon satellites had their orbits perturbed by the pressure of sunlight itself. This force becomes important for objects with a very large surface area to mass ratio, and solar sailing may become a viable propulsion technique in the near future.

4.2 The Gravitational Field Equation

Gravity is a conservative force. For any distribution of mass, the acceleration of a test particle can be expressed as

$$\mathbf{a}_g = -\nabla V(x, y, z) \tag{4.1}$$

where $\mathbf{a}_g$ is the gravitational acceleration of the test particle, ∇ is the gradient operator, and $V(x, y, z)$ is the potential energy per unit mass. We again follow the convention of working with the specific potential, since the mass of the satellite may be unknown. We know how to write the potential V for a Newtonian point mass, but we do not know how, as yet, to write the potential for a nonspherical distribution of mass. Since planets are generally not perfect spheres, we need to study more general potential functions.

Begin with the expression for the gravitational acceleration created by a single point mass m at the origin, as shown in Figure 4.1. This is given by the familiar

$$\mathbf{a}_g = -\frac{Gm\mathbf{r}}{r^3} \tag{4.2}$$

Now, surround the mass with a closed surface S, and let us integrate the normal component of the acceleration over this surface

$$\oint_S \mathbf{a}_g \cdot \mathbf{n} da = -Gm \oint_S \frac{\mathbf{r} \cdot \mathbf{n}}{r^3} da \tag{4.3}$$

where $\mathbf{n}$ is the surface normal vector, and da is the differential surface area. Essentially, we are trying to calculate how much total gravitational field the point mass creates.

Now, the integral above is independent of the size and shape of the surface S, so long as S is a simple convex surface. (We have not proved this!) So, let us replace the general surface S with a unit sphere. For such a sphere, $r = 1, \mathbf{r} = \mathbf{n}$, the surface normal vector, and $da = d\Omega$, the element of solid angle. Since there are 4π steradians in a sphere, the right side of (4.3) integrates to

$$\oint_S \mathbf{a}_g \cdot \mathbf{n} da = -4\pi Gm \tag{4.4}$$

Now, on the left side of the above, we may appeal to the divergence theorem to replace the surface integral with a volume integral

$$\oint_S \mathbf{a}_g \cdot \mathbf{n} da = \int_v \nabla \cdot \mathbf{a}_g \, dv \tag{4.5}$$

where dv is the differential volume element. Also, the right side of (4.4), $4\pi Gm$, can also be replaced with a volume integral over the mass density function ρ within the surface. This step removes the single point mass assumption, and replaces it with a general mass distribution

$$m = \int_v \rho \, dv \tag{4.6}$$

We thus arrive at the statement

$$\int_v \nabla \cdot \mathbf{a}_g \, dv = -4\pi G \int_v \rho \, dv \tag{4.7}$$

in terms of two volume integrals.

Now, the above equation is true independent of the size and shape of the volume v. In this case it is permissible to equate the integrands, obtaining

$$\nabla \cdot \mathbf{a}_g = -4\pi G\rho \tag{4.8}$$

Finally, since the acceleration can be written as minus the gradient of a potential function $V(x, y, z)$, we obtain

$$\nabla \cdot \nabla V = 4\pi G\rho \tag{4.9}$$

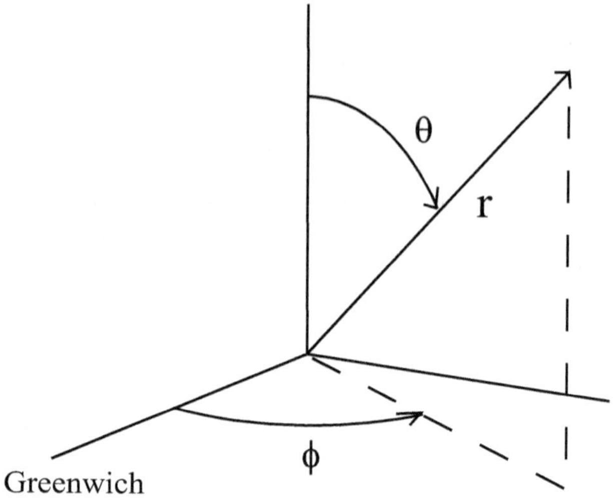

Figure 4.2: Coordinates for Geopotential Calculation

The operator $\nabla \cdot \nabla$ is usually written ∇^2, and is called the *Laplacian operator.* Equation (4.9) thus assumes the final form

$$\nabla^2 V = 4\pi G\rho \tag{4.10}$$

This is called *Poisson's equation.* It is the fundamental partial differential equation for gravitational fields. That is, if the mass distribution $\rho(x, y, z)$ is known, equation (4.10) can (in principle) be solved to give the gravitational potential function $V(x, y, z)$.

4.3 The Geopotential

There is no closed form solution to the general case of Poisson's equation. However, most of the objects that a satellite might decide to orbit are basically spherical. Also, we are only interested in the gravitational potential outside of this sphere. Since the satellite will be orbiting within a vacuum, Poisson's equation simplifies to the statement

$$\nabla^2 V = 0 \tag{4.11}$$

since $\rho = 0$. Assume that the equatorial radius of the earth is given by $R_\oplus$. Since the equatorial radius is the largest radius of the earth, a sphere with this radius will contain all of the mass of the earth. As shown in Figure 4.2, introduce spherical polar coordinates r, θ, ϕ. Also, since the mass distribution irregularities rotate with the rotation of the earth, we will assume that the coordinate frame in the figure rotates with the earth, and therefore ϕ is simply the geocentric longitude, while θ is the colatitude.

Now, in spherical polar coordinates, with zero mass density, Poisson's equation assumes the form

$$\begin{aligned}\frac{1}{r^2}\frac{\partial}{\partial r}\left(r^2\frac{\partial V}{\partial r}\right) &+ \frac{1}{r^2\sin\theta}\frac{\partial}{\partial\theta}\left(\sin\theta\frac{\partial V}{\partial\theta}\right) \\ &+ \frac{1}{r^2\sin^2\theta}\frac{\partial^2 V}{\partial\phi^2} = 0\end{aligned} \tag{4.12}$$

Although this is formidable in appearance, it is a linear partial differential equation. That is, if V_1 and V_2 are two solutions to (4.12), then

$$V = c_1 V_1 + c_2 V_2 \tag{4.13}$$

is also a solution, where c_1 and c_2 are arbitrary constants. This equation separates if we assume that

$$V(r, \theta, \phi) = R(r)\Theta(\theta)\Phi(\phi) \tag{4.14}$$

When this is inserted into (4.12), and multiplied by r^2, the result is

$$\begin{aligned}\Theta\Phi\frac{d}{dr}\left(r^2\frac{dR}{dr}\right) &+ R\Phi\frac{1}{\sin\theta}\frac{d}{d\theta}\left(\sin\theta\frac{d\Theta}{d\theta}\right) \\ &+ R\Theta\frac{1}{\sin^2\theta}\frac{d^2\Phi}{d\phi^2} = 0\end{aligned} \tag{4.15}$$

Since each of the functions R, Θ, Φ, are functions of one variable, the partial derivatives in (4.12) have been replaced by total derivatives.

To perform the first separation, multiply by $\sin^2\theta / R\Theta\Phi$, and rewrite the equation as

$$\begin{aligned}\frac{\sin^2\theta}{R}\left(\frac{d}{dr}\left(r^2\frac{dR}{dr}\right)\right) &+ \frac{\sin\theta}{\Theta}\left(\frac{d}{d\theta}\left(\sin\theta\frac{d\Theta}{d\theta}\right)\right) \\ &= -\frac{1}{\Phi}\frac{d^2\Phi}{d\phi^2}\end{aligned} \tag{4.16}$$

Now, the left side is a function of r and θ, while the right side is a function of ϕ only. This must be true at any point (r, θ, ϕ) outside the radius of the earth. The only way that a function of r and θ alone can equal a function of ϕ alone for any values of r, θ, ϕ, is for both sides to not really be functions of their indicated variables. Instead, both sides must be constant. Assume for the moment that this constant is k.

Equating the right side of (4.16) to k, we obtain

$$k = -\frac{1}{\Phi}\frac{d^2\Phi}{d\phi^2} \tag{4.17}$$

This can be rearranged to give

$$\frac{d^2\Phi}{d\phi^2} = -k\Phi \tag{4.18}$$

This is the equation for a simple harmonic oscillator. Its general solution is $\Phi = C\cos\sqrt{k}\phi + S\sin\sqrt{k}\phi$, where C and S are constants. However, not every value of k can be permitted in this solution. If we circle the equator once, the function Φ must return to its starting value. If it does not, then there will be a jump discontinuity in the potential V, and this will lead to infinite gravitational accelerations when we calculate the gradient of the potential. Instead, we must impose the boundary condition $\Phi(0) = \Phi(2\pi)$. This requires that $\sqrt{k} = m$, where m is a non – negative integer, and the general solution for the Φ function becomes

$$\Phi_m = C_m \cos m\phi + S_m \sin m\phi \tag{4.19}$$

Returning to the remainder of Poisson's equation, we have

$$\frac{\sin^2\theta}{R}\left(\frac{d}{dr}\left(r^2\frac{dR}{dr}\right)\right) + \frac{\sin\theta}{\Theta}\left(\frac{d}{d\theta}\left(\sin\theta\frac{d\Theta}{d\theta}\right)\right) = m^2 \tag{4.20}$$

Dividing by $\sin^2\theta$, the second separation can be effected

$$\frac{1}{R}\frac{d}{dr}\left(r^2\frac{dR}{dr}\right) = l = -\frac{1}{\Theta\sin\theta}\frac{d}{d\theta}\left(\sin\theta\frac{d\Theta}{d\theta}\right) + \frac{m^2}{\sin^2\theta} \tag{4.21}$$

where the separation constant is l. Again, this depends on the argument that a function of r (the left side) can only equal a function of θ (the right side) everywhere only if both are not really functions of what they claim to be.

The right side can be rewritten as

$$\frac{1}{\sin\theta}\frac{d}{d\theta}\left(\sin\theta\frac{d\Theta}{d\theta}\right) - \left(\frac{m^2}{\sin^2\theta} - l\right)\Theta = 0 \tag{4.22}$$

This is called *Legendre's equation.* Its solution is discussed in any book on special functions (for example, Handbook of Mathematical Functions, Abramowitz and Segun). If we were to change independent variables to $x = \cos\theta$, we would obtain a form of Legendre's equation which would have polynomial solutions in x.

Again, there is a boundary condition on the permissible solutions to (4.22), this time we must require that there be no slope discontinuity in the function Θ at the poles:

$$\left.\frac{d\Theta}{d\theta}\right|_{\theta=0 \ or \ \pi} = 0 \tag{4.23}$$

since this will also lead to infinite gravitational accelerations. The solutions to (4.22) are called the *associated Legendre polynomials,*

$$\Theta(\theta) = P_n^m(\cos\theta) \tag{4.24}$$

which is a polynomial in $\cos\theta$. The boundary condition also leads to the restriction that $l = n(n+1)$, where n is a positive integer.

Finally, the remainder of Poisson's equation is

$$\frac{d}{dr}\left(r^2\frac{dR}{dr}\right) = n(n+1)R \tag{4.25}$$

If we insert the trial solution $R = r^p$, we find that permissible values of p are $p = n$ and $p = -(n+1)$. That is, the two possible solutions are

$$R = r^n, \quad R = r^{-n-1} \tag{4.26}$$

The first is unacceptable physically, since the potential increases without limit with r. (This would be the appropriate solution function *inside* our sphere of radius $R_\oplus$, however, if the sphere was empty and all the mass was outside.) The correct form is thus the second one, $R = r^{-n-1}$.

Finally, any combination of (4.19), (4.24), and (4.26) may be multiplied together, so long as the values of n and m match. Also, any combination of $R(r)\Theta(\theta)\Phi(\phi)$ may then be added together. The result is the general infinite series expansion for the potential function outside of a sphere of radius $R_\oplus$:

$$\begin{aligned} V(r,\theta,\phi) &= -\frac{\mu}{r}\sum_{n=0}^{\infty}\sum_{m=0}^{n}\left(\frac{r}{R_\oplus}\right)^{-n} P_n^m(\cos\theta) \\ &\times \left(C_{nm}\cos m\phi + S_{nm}\sin m\phi\right) \end{aligned} \tag{4.27}$$

This is called the expansion of the geopotential in spherical harmonics, or just the geopotential expansion for short. The constants C_{nm} and S_{nm} have been redefined so that our familiar friend

$$V = -\frac{\mu}{r} \tag{4.28}$$

appears as the first term. Since $P_0^0 = 1$, this means that $C_{00} = 1$ and $S_{00} = 0$. The other C_{nm} and S_{nm} specify the actual shape of the gravitational field. The divisions by the equatorial radius of the earth ($R_\oplus$) have been chosen to make the C_{mn} and S_{mn} dimensionless numbers. A table of these values is called a gravity model, several of which are available.

4.4 Satellite Equations of Motion

As a first example of the use of the geopotential expansion, consider the motion of a satellite. In section 2.9 we solved the two body problem in spherical polar coordinates very similar to the set used in the previous section. The only difference is that in section 2.9 ϕ was the right ascension of the satellite, while ϕ in the last section is the longitude. Since the right ascension is related to the longitude by

$$\phi_{long} = \phi_{RA} - \omega_\oplus t \tag{4.29}$$

it is now a simple matter to obtain the equations of motion for an earth satellite.

We need only extract the kinetic energy part of the Hamiltonian function from section 2.9, and replace the potential energy of the two body problem with the full geopotential expansion to obtain

$$\begin{aligned} H &= \frac{1}{2}\left(p_r^2 + \frac{1}{r^2}p_\theta^2 + \frac{1}{r^2\sin^2\theta}p_\phi^2\right) \\ &- \frac{\mu}{r}\sum_{n=0}^{\infty}\sum_{m=0}^{n}\left(\frac{r}{R_\oplus}\right)^{-n} P_n^m(\cos\theta) \\ &\times \left(C_{nm}\cos m(\phi - \omega_\oplus t) + S_{nm}\sin m(\phi - \omega_\oplus t)\right) \end{aligned} \tag{4.30}$$

where the ϕ retained is the right ascension of the satellite.

The equations of motion can now be obtained from Hamilton's equations

$$\dot{r} = \frac{\partial H}{\partial p_r} = p_r \tag{4.31}$$

$$\dot{\theta} = \frac{\partial H}{\partial p_\theta} = \frac{1}{r^2}p_\theta \tag{4.32}$$

$$\dot{\phi} = \frac{\partial H}{\partial p_\phi} = \frac{1}{r^2\sin^2\theta}p_\phi \tag{4.33}$$

$$\dot{p}_r = -\frac{\partial H}{\partial r} = \frac{1}{r^3}\left(p_\theta^2 + \frac{1}{\sin^2\theta}p_\phi^2\right) \tag{4.34}$$

$$\begin{aligned} &- \mu\sum_n\sum_m (n+1)\frac{R_\oplus^n}{r^{n+2}}P_n^m(\cos\theta) \\ &\times \left(C_{nm}\cos m(\phi - \omega_\oplus t) + S_{nm}\sin m(\phi - \omega_\oplus t)\right) \end{aligned} \tag{4.35}$$

$$\begin{aligned} \dot{p}_\theta &= -\frac{\partial H}{\partial \theta} = \frac{\cos\theta}{\sin^3\theta}p_\phi^2 + \frac{\mu}{r}\sum_n\sum_m\left(\frac{r}{R_\oplus}\right)^{-n} \\ &\times \frac{d}{d\theta}\left(P_n^m(\cos\theta)\right)\left(C_{nm}\cos m(\phi - \omega_\oplus t)\right. \\ &+ \left. S_{nm}\sin m(\phi - \omega_\oplus t)\right) \end{aligned} \tag{4.36}$$

$$\begin{aligned} \dot{p}_\phi &= -\frac{\partial H}{\partial \phi} = -\frac{\mu}{r}\sum_n\sum_m\left(\frac{r}{R_\oplus}\right)^{-n} P_n^m(\cos\theta) \\ &\times m\left(-C_{nm}\sin m(\phi - \omega_\oplus t) + S_{nm}\cos m(\phi - \omega_\oplus t)\right) \end{aligned} \tag{4.37}$$

The partial derivatives of the potential function are not terribly difficult to obtain, since the dependence on r and ϕ is explicit, while the Legendre functions are known functions of $\cos\theta$. So, in this (or analogous) forms the equations of motion can be numerically integrated, at worst, to find the orbit. However, this is exceedingly expensive if not done carefully. The topic of recursion relations for these functions will be taken up in the next section.

Even with this numerical streamlining, the full equations of motion can be very expensive to numerically integrate, especially if the time interval to be covered is long. Since numerical integration must get from t_1 to t_2 by going "through" every moment in between, it may not be as efficient as a general perturbation solution, which can go to t_2 in one "step". However, the series solution approach does not easily handle air drag, and is much more involved to program. Generally, a numerical integration is called for when the time interval is short, and the utmost in accuracy is needed. As computers have become cheaper and more prevalent, numerical integration has becoming the de–facto standard way to propagate an orbit.

Finally, note that the Hamiltonian (4.31) is a function of all the variables, and explicitly a function of time. It appears to have no integrals of the motion. But if we pose this problem in a reference frame that rotates with the earth, this situation will change, and there will be an integral. This is left as a problem.

4.5 Learning to Love the Geopotential

In this section we will spend some time getting acquainted with the individual terms in the geopotential expansion. Since they have different effects on an orbit, and are of different relative importance, this will give us some familiarity with the terminology used for these different terms. Also, we will be able to picture their individual contributions to the total gravitational field.

We have already arranged for the Newtonian point mass potential to appear as the first ($n = 0, m = 0$) term in the geopotential expansion. Since $m = 0$, $\cos m\phi = 1$, $\sin m\phi = 0$, and since the first Legendre polynomial $P_0^0 = 1$, the first term does not depend on either the longitude or the colatitude. With $C_{00} = 1$ and $S_{00} = 0$ (actually this latter condition is not needed), the lead term returns the Newtonian potential.

The correct units on the potential are assured by the presence of μ in front of the expansion, and by dividing the higher order dependence on the radius r by the radius of the earth $R_\oplus$. This makes all of the coefficients C_{nm} and S_{nm} dimensionless. Since the Newtonian term is represented by $C_{00} = 1$, the magnitude of higher order C and S coefficients is one indicator of their relative importance. The C_{n0} coefficients are also written $-J_n$, while all S_{n0} coefficients are unnecessary.

While the form (4.27) seems to be in most common use among those orbital specialists who actually *use* the geopotential, there is another form used by those who *determine* the geopotential. The use of "fully normalized" Legendre polynomials is common when one actually searches out the gravity models. These are defined as

$$\bar{P}_n^m(z) = \left[\frac{\epsilon_m(2n+1)(n-m)!}{(n+m)!}\right]^{1/2} P_n^m(z) \tag{4.38}$$

where

$$\epsilon_m = \begin{cases} 1 & m = 0 \\ 2 & m \neq 0 \end{cases} \tag{4.39}$$

This is really just a change in the normalization of the Legendre polynomials, and the inverse change is made in the gravity coefficients C_{nm}, S_{nm}. Most gravity models are now cited in terms of these fully normalized gravity harmonic coefficients. A quick way to tell which set of coefficients is being cited is to find the familiar C_{20}, and determine if it is nearly the familiar (not fully normalized) $C_{20} = -J_2 \approx -0.001082$, or not. If multiplying by $\sqrt{5}$ produces the familiar value, the set of coefficients is of the fully normalized variety. It is perhaps simplest to do a one-time conversion of the fully normalized coefficients $\bar{C}_{nm}$, $\bar{S}_{nm}$ back to the unnormalized form.

The geopotential expansion of the earlier section explicitly shows its dependence on radius and longitude, but the latitude dependence is tied up in the associated Legendre polynomials $P_n^m(\cos\theta)$. The first few of these are given by

$$\begin{aligned}
P_0^0 &= 1, \quad P_1^0 = \cos\theta \\
P_2^0 &= \frac{1}{2}(3\cos^2\theta - 1) \\
P_3^0 &= \frac{1}{2}(5\cos^3\theta - 3\cos\theta) \\
P_4^0 &= \frac{1}{8}(35\cos^4\theta - 30\cos^2\theta + 3) \\
P_5^0 &= \frac{1}{8}(63\cos^5\theta - 70\cos^3\theta + 15\cos\theta)
\end{aligned}$$

which are all for order n and degree $m = 0$.

The above table can be extended by several means. Rodrigues' formula

$$P_n^0(\cos\theta) = \frac{1}{2^n n!}\frac{d^n}{d(\cos\theta)^n}\left(\cos^2\theta - 1\right)^n \tag{4.40}$$

will give all of the zero degree Legendre polynomials. The associated (non - zero m) polynomials can be found from

$$P_n^m(\cos\theta) = (-1)^m \sin^m\theta \frac{d^n}{d(\cos\theta)^n} P_n^0(\cos\theta) \tag{4.41}$$

However, these produce the expressions for the Legendre polynomials. Individually programming dozens of individual polynomials and their derivatives is not an inviting task, and will almost certainly result in programming errors. There is a better way to handle this problem.

The geopotential equations of motion are doubly infinite series expansions. Although this cannot be avoided, there are certain steps which must be taken to ease the computational burden of evaluating the equations of motion. Let us begin with the easy and easily overlooked terms. The sine and cosine functions are usually quite expensive to calculate within the computer, and many of them are required to evaluate the equations of motion at each time step. However, if we use the trigonometric identities

$$\begin{aligned}\sin(m+1)\phi &= \sin\phi\cos m\phi + \cos\phi\sin m\phi \\ \cos(m+1)\phi &= \cos\phi\cos m\phi - \sin\phi\sin m\phi\end{aligned}$$

as *recursion relations,* we can build a table of values of $\sin m\phi$ and $\cos m\phi$ for the cost of only two multiplications each. Similarly, the seemingly trivial

$$r^{-n-1} = \left(r^{-n}\right)\left(r^{-1}\right) \tag{4.42}$$

allows us to avoid computer calculation of an exponent. This usually involves calculating a logarithm, multiplying by the exponent, and exponentiating the result. This is an expensive process numerically, the simple multiplication in the above (multiply by r^{-1}, not divide by r!) is much cheaper. There are similar recursion relations for the Legendre polynomials. These should be used to similarly ease the computational burden, as well as the programming burden of coding dozens of the $P_n^m(\cos\theta)$. To be specific, with $z = \cos\theta$:

$$P_n^{m+1}(z) = \frac{(n-m)zP_n^m(z) - (n+m)P_{n-1}^m(z)}{\sqrt{1-z^2}} \tag{4.43}$$

allows us to increase the order (m) of the polynomial, while

$$P_{n+1}^m(z) = \frac{(2n+1)zP_n^m(z) - (n+m)P_{n-1}^m}{n-m+1} \tag{4.44}$$

permits changing the degree (n) of the polynomial. In fact, a little checking will show that $P_0^0 = 1$ and $P_1^0 = \cos\theta$ are the only two Legendre polynomials that need be calculated explicitly. All of the rest can be obtained from the above. These forms of the Legendre polynomial recursions become numerically unstable when near the poles, where $z^2 \to 1$. Then it is necessary to begin at the other end, and starting from

$$P_n^n(z) = \frac{(-1)^n(2n)!}{2^n n!}z^n, \quad P_n^{n-1}(z) = -\frac{zP_n^n(z)}{\sqrt{1-z^2}} \tag{4.45}$$

and then use (4.43) as a *downward* recusion relationship in m, with the dangerous quantity z^2-1 expanded in a binomial series when $z \approx 1$.

Similarly, the derivatives of the Legendre polynomials

$$\frac{d}{dz}P_n^m(z) = \frac{nzP_n^m(z) - (n+m)P_{n-1}^m(z)}{z^2-1} \tag{4.46}$$

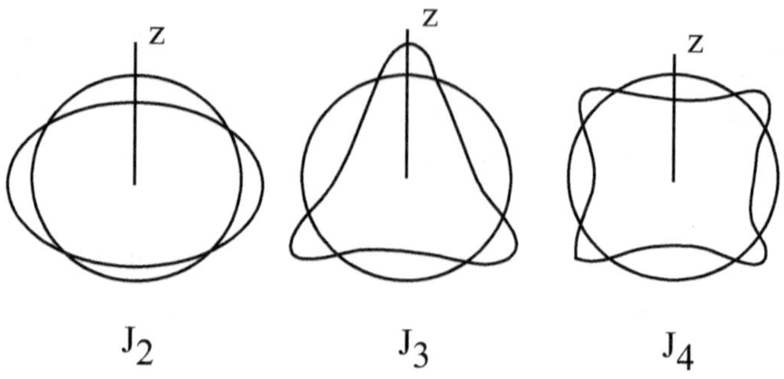

Figure 4.3: The First Three Zonal Harmonics

can also be calculated without individual programming. These derivatives are needed for the equations of motion. Higher order derivatives obey relations which can be found by calculating derivatives of any of the recursion relations given here. As the code for evaluating the geopotential equations of motion is buried deep within the numerical integration, any options to speed up the calculations should be exercised. Some of the above injunctions are not operative on computers with high speed numerical coprocessors, but a system programmer cannot always depend on this being the case.

We will now spend some time getting familiar with exactly what the individual terms model in the earth's gravitational field. The terms in the geopotential expansion for degree $m = 0$ are called *zonal harmonics.* The first zonal harmonic, for $n = 1$ and $m = 0$ shifts the center of mass of the entire earth north or south along our z axis. It is the only term which can do this, and we have little interest in a coordinate frame whose origin is not at the center of mass. Choosing as origin the center of mass of the earth thus makes $J_1 = 0$, and the first zonal harmonic vanishes. The first non-vanishing zonal harmonic is thus the $n = 2$ term. It contributes a term to the potential which looks like

$$V_{20} = \frac{\mu R_\oplus^2 J_2}{2r^3} \left(3\cos^2\theta - 1\right) \tag{4.47}$$

For the earth, the coefficient $J_2 = -C_{20} = 0.001082$, which is about 10^{-3} of the contribution of the Newtonian point mass term. Physically, this term represents the earth's equatorial bulge. The Legendre polynomial is positive in a zone near the earth's equator, and negative in two zones near the poles. This leads to higher gravitational potential over the equatorial zone, a consequence of the additional mass in this region. This is the dominant term in the earth's potential expansion after the Newtonian term, and for all other planets as well.

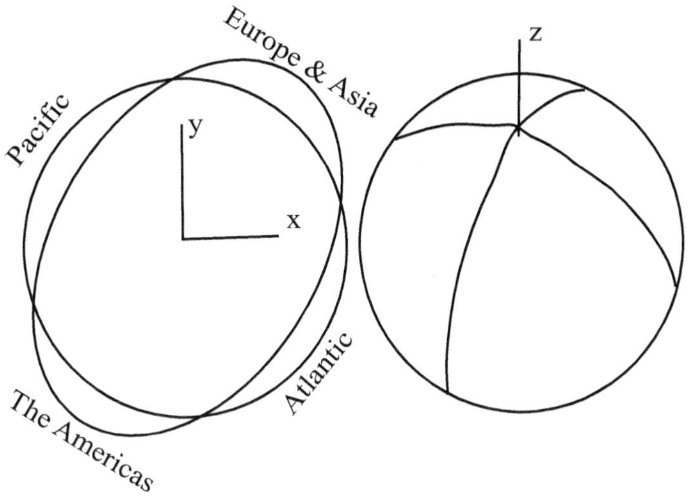

Figure 4.4: The C_{22}, S_{22} Sectoral Harmonic

Rotation causes planets to bulge at the equator, and this leads to relatively large J_2 terms.

We can picture the contribution to the potential as shown in Figure 4.3. The sketches show both the rough shape of the earth which causes the additional potential, as well as the bands of higher and lower potential, for the first three zonal harmonics. A perfectly fluid planet (and the gas giant planets Jupiter, Saturn, Uranus, and Neptune are such) would be symmetric about their equator planes, and would not have any structure in their longitude directions. The only terms expected in their gravitational fields are thus the even order ($n = 2, 4, 6, ...$) zonal harmonics. Note that the higher order terms fall off as r^{-n-1} in the potential, which means that the force terms they generate fall off like r^{-n-2}. High order terms are thus only significant in the immediate vicinity of the planet, and at large distances the geopotential expansion reverts to its first term, the Newtonian point mass potential.

At the opposite extreme are the terms for which $n = m$. From the definition of the associated Legendre polynomials, we see that P_n^n is proportional to $\sin^n \theta$. This expression does not have any zeros in the colatitude, so these terms do not have any sign changes in the latitude. Instead, they are functions of longitude only, and are called *sectoral harmonics.* The $n = m = 1$ sectoral harmonic has the effect of moving the earth's center of mass off the origin of our coordinate frame to some other point within the plane of the equator. Again, this is not desirable, so the choice $S_{11} = C_{11} = 0$ is made. The first non - vanishing

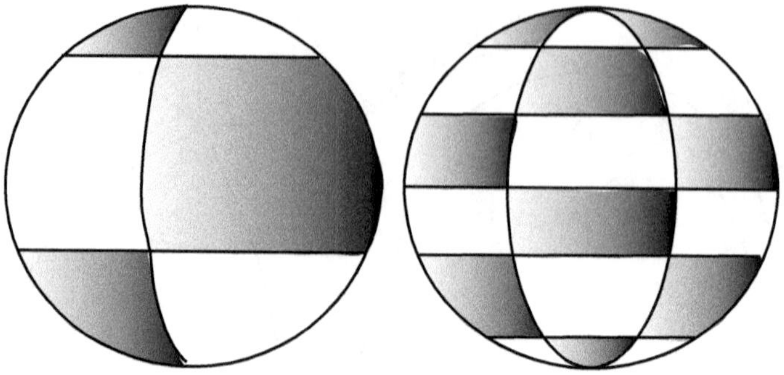

Figure 4.5: Representative Tesseral Harmonics

sectoral harmonic is thus the $n = m = 2$ term, which is proportional to

$$V_{22} \propto C_{22} \cos 2\phi + S_{22} \sin 2\phi$$

where ϕ is again the longitude. This generates two regions of high potential, with two regions of lower potential between them, arranged like the sectors of an orange. The largest sectoral harmonic for the earth is the $n = m = 2$ term shown in Figure 4.4, with the regions of high potential over the two large continental blocks (Eurasia and the Americas) and the two regions of low potential aligned with the two large ocean basins (Atlantic and Pacific). The earth seems to have an unusually large asymmetry in this regard, yet the coefficients C_{22} and S_{22} are about 10^{-6}, or three orders of magnitude smaller than J_2 .

The general case when $m \neq 0$ and $n \neq m$, is called a *tesseral harmonic.* It combines the latitude dependent banding of the zonal harmonics with the longitude dependence of the sectoral harmonics to produce a checkerboard of high and low potential regions on the sphere. Since the general associated Legendre polynomial has $n - m$ zeros in the interval $0 < \theta < \pi$, the sphere will have $n - m + 1$ bands in latitude. There are still $2m$ divisions in longitude due to the $\sin m\phi$, $\cos m\phi$ dependence. Some tesseral harmonics are shown in Figure 4.5. The word comes from the Latin *tesserae,* or "tiles". The radial dependence of high order terms again drops off greatly with even a slight increase in distance.

A table of the coefficients C_{nm} and S_{nm} is called a *gravity model.* Several are available, including The Smithsonian Standard Earth III (1971) and the World Geodetic Survey series of models. Known by their acronymns as SAO3 and WGS, they are relatively complete through $n, m < 25$. Recently, these have

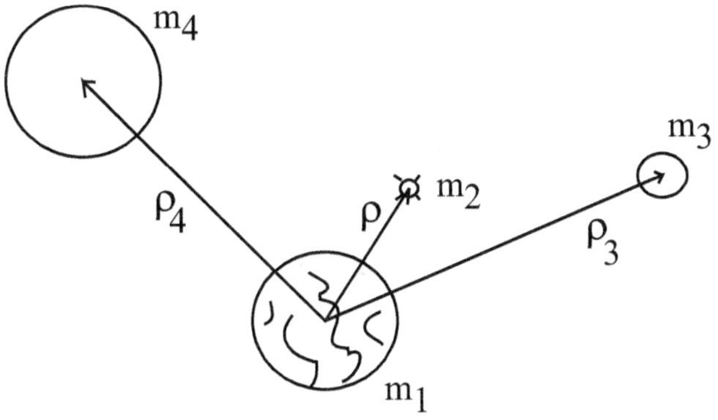

Figure 4.6: Relative Position Vectors

been superseded by the NASA Earth Gravity Model 1996, or EGM96, which can be found on the world wide web. At least some terms are available for most planets in the solar system, and relatively complete gravitational models are available for the moon, Mars, Venus, Jupiter, and Saturn, besides, of course, the earth.

4.6 Third Body Gravitational Forces

When air drag and higher order geopotential forces are added to the equations of motion, the underlying reference frame remains inertial. This is not true when a "third" gravitating body is added to the system, since its gravitational field will influence both the motion of the satellite and the motion of the earth. With our usual techniques, this would force us to set up the problem anew, using a true inertial reference frame, and the kinetic energy part of the Hamiltonian would have to be rederived. This is not nearly so convenient as simply adding or removing force/potential terms as necessary to describe the system. However, there is a formulation of third body gravitational effects which enables us to retain the advantages of a potential energy formulation, without having to reconstruct the kinematics.

As shown in Figure 4.6, let us start with the equations of motion for the N body problem with respect to a true inertial reference frame

$$\ddot{\mathbf{r}}_i = -\sum_{j \neq i}^{n} \frac{\mu_j}{r_{ij}^3}(\mathbf{r}_i - \mathbf{r}_j) \tag{4.48}$$

where $r_{ij} = |\mathbf{r}_i - \mathbf{r}_j|$ and $\mu_j = Gm_j$. Now, suppose that $i = 1$ is the earth, or the object we are orbiting. Also, let body $i = 2$ be the object for which we wish to write equations of motion. If we let the relative position vector be $\vec{\rho}_j = \mathbf{r}_j - \mathbf{r}_1$, and $\rho_{ij} = |\mathbf{r}_i - \mathbf{r}_1 - \mathbf{r}_j + \mathbf{r}_1| = r_{ij}$, then the equations of motion for $\vec{\rho}_{12} = \rho$, can be easily found, since

$$\ddot{\vec{\rho}} = \ddot{\mathbf{r}}_2 - \ddot{\mathbf{r}}_1 \tag{4.49}$$

Substituting for $\ddot{\mathbf{r}}_1$ and $\ddot{\mathbf{r}}_2$, we have

$$\ddot{\vec{\rho}} = -\mu\frac{\vec{\rho}}{\rho^3} - \sum_{j=3}^{n}\mu_j\left(\frac{\vec{\rho}-\vec{\rho}_j}{\rho_{2j}^3} + \frac{\vec{\rho}_j}{\rho_j^3}\right) \tag{4.50}$$

where $\mu = \mu_1 + \mu_2$. The first term is obviously the two body term we expected. In the summation, the first term is the direct action of the third body on the satellite. That is, it is what we would have added to the equations of motion if we forgot that the third body makes the earth – centered reference frame noninertial. The last term is called the indirect term, and reflects perturbations in the motion of the satellite because the third body has perturbed the earth itself.

However, to be really useful, we would like a potential energy formulation of the above. With a potential energy, it is simple to incorporate a new force into a system Hamiltonian. We would only need to convert the potential energy into the coordinates and momenta used in the problem, and then take the appropriate partial derivatives. This is also true for the potential form of the Lagrange planetary equations. So, we need to see if the perturbing terms in (4.50) can be obtained as the gradient of some potential function.

The direct term is the most obvious. Since it represents the direct gravitational attraction of the third mass on the satellite, it seems natural to try

$$V_{direct} = -\sum_{j=3}^{n}\frac{\mu_j}{\rho_{2j}} \tag{4.51}$$

as the potential. Calculation of the gradient of this with respect to the $\vec{\rho}$ vector (the coordinates of the satellite) shows that this is indeed the potential energy of the direct third body – satellite interaction.

The indirect term, however, represents the kinematic effects of the noninertial origin at the center of the earth. It is less obvious that this might be derivable from a potential energy. However, the gradient must be taken with respect to the coordinates of the satellite, and the indirect term does not contain the coordinates of the satellite at all. So, the potential for the indirect part is given by

$$V_{indirect} = +\sum_{j=3}^{n}\mu_j\frac{\vec{\rho}_j\cdot\vec{\rho}}{\rho_j^3} \tag{4.52}$$

Calculation of the gradient with respect to $\vec{\rho}$ then simply returns the terms we needed.

So, the complete interaction of a third gravitating object can be represented by the potential function

$$V = -\sum_{j=3}^{n} \mu_j \left(\frac{1}{\rho_{2j}} - \frac{\vec{\rho}_j \cdot \vec{\rho}}{\rho_j^3} \right) \tag{4.53}$$

which includes the effects of the motion of the earth itself caused by this object. If we are working with the Lagrange Planetary equations, the disturbing function $R = -V$. For the motion of an earth satellite, the most obvious "third" bodies are the sun and moon. Since the satellite itself will be too small to influence their motion, the position vectors of the sun and moon with respect to the earth can be treated as known functions of time (meaning that someone else has already calculated their orbits). While (4.53) is so widely used in solar system dynamics that it has earned the title "The Disturbing Function", it is not really a true potential energy. Calculation of partial derivatives with respect to the coordinates of body j will not correctly give the force of the satellite on body j. This is not a difficulty if the satellite is too small to influence the motion of body j. A true potential energy would give both the force of body j on body 2, *and* the force of body 2 on body j. Another disturbing function must be written if the influence of body "2" on body j is to be studied.

4.7 Air Drag

Satellites encounter air molecules at very high speeds, and at densities so low that the air no longer behaves like a fluid. Instead, the air molecules behave as individual particles, in what is called the free molecular flow regime. If the air density is ρ, then the number of molecules hit per second by the satellite is proportional to ρV, where V is the speed of the satellite with respect to the air. Now, force is the rate of change of momentum, and as each air molecule hits and bounces off, it transfers linear momentum (through inelastic collision) proportional to V. The air drag force is thus proportional to $\rho V^2/2$, which is called the dynamic pressure. Also, the force will be proportional to the presented area of the satellite, A. However, we have made no allowance for molecules which reflect obliquely off the surface. The drag thus depends on an effective area $C_D A$, where the coefficient of drag, or ballistic coefficient, C_D, is the slop term needed to correct for oblique reflections due to the shape of the object. Finally, drag is by definition directed opposite the velocity vector (any other aerodynamic force component is lift), and we wish to work with the acceleration due to drag, not the force itself. The drag law has the form

$$\mathbf{a}_d = -\frac{\mathbf{V}}{V} \frac{C_D A}{m} \frac{1}{2} \rho V^2 = -\frac{1}{2} \frac{C_D A}{m} \rho V \mathbf{V} \tag{4.54}$$

This form of the drag law was first derived by Isaac Newton himself, who was under the impression that a fluid was a group of noninteracting particles. It is still a correct description of drag when the fluid is so diffuse that the particles don't interact, and when the velocity of the vehicle is much greater than the speed of sound in the fluid.

As mentioned in a earlier chapter, the coefficient of drag, C_D, the area A, and the satellite mass m may not be separately determinable. The first two quantities do not appear anywhere else within the equations of motion for orbital systems, and the mass m can normally be divided out of those equations of motion. So, the one quantity

$$B^* = \frac{C_D A}{m} \tag{4.55}$$

is usually all that can be determined, and the drag law can be rewritten as

$$\mathbf{a}_d = -\frac{1}{2} B^* \rho V \mathbf{V} \tag{4.56}$$

There are a few problems with dealing with air drag on a satellite. First, the velocity V in (4.56) is the velocity of the satellite with respect to the atmosphere. To a good approximation, the atmosphere rotates with the earth as if it was a solid body. So, neglecting the local wind vector, the relative velocity $\mathbf{V}$ is

$$\mathbf{V} = \mathbf{v}_{in} - \omega_{\oplus} \times \mathbf{r} \tag{4.57}$$

where $\mathbf{v}_{in}$ is the satellite's inertial velocity, and $\mathbf{r}$ is its inertial position vector. The rotational velocity of the atmosphere can be as much as 1/2 km/sec near the equator, so this is not a totally negligible correction. For a polar orbit it would represent the largest out–of–plane force.

However, this is a simple deterministic correction. The real difficulties arise from the innocent looking B^* and ρ terms. A satellite's effective area to mass ratio B^* involves it's shape and rotational state, and these may change with time. Also, during the final plunge into the atmosphere, we have no guarantees whatever that the shape, rotational state, and even the number of component parts will not change with time.

Even for a simple spherical satellite not undergoing final decay, the atmospheric density will make the prediction of air drag effects a very uncertain process. There are predictable effects where the atmosphere is known to bulge over the equator (the equatorial bulge), and due to heating effects, the atmosphere bulges over the mid - afternoon longitudes (the diurnal bulge). The diurnal bulge also follows the sun during the course of the year, staying close to the subsolar point on the earth. These effects are rather stable and predictable.

There are, however, effects at high altitude which are very unpredictable. The atmosphere can be thought of as consisting of three layers: the troposphere, the stratosphere, and the exosphere. The troposphere is dominated by heating from the surface, is usually in convection during the day, and contains all of the unpredictable weather we suffer here on the surface. The stratosphere lies above the troposphere, and is a stable, basically unchanging region. The exosphere is heated directly by interaction with the earth's radiation belts, and this mechanism is just as unpredictable as the occurrence of thunderstorms at lower altitudes.

The heating of the exosphere rises and falls on an 11 year cycle, in concert with the 11 year sunspot cycle. The rise of solar activity feeds more charged particles into the earth's radiation belts. As these leak out of the radiation belts near the earth's poles, this leads to a general expansion of the earth's atmosphere during sunspot maximum, and a contraction during sunspot minimum. This effect is also predictable. However, a sudden catastrophic release of energy on the sun (a solar flare) can dump huge amounts of energetic particles into the earth's radiation belts, and these immediately begin to leak into the exosphere, causing it to heat and expand. This will lead to a sudden increase in air drag on all low earth orbiting satellites.

An analogous effect occurs when the magnetic field of the earth suddenly changes. The earth's magnetic field is not static, but can undergo moderate changes on a short timescale. (Iron becomes nonmagnetic at about 500°, the Curie point. The earth's magnetic field is generated by a self – excited dynamo effect.) This can have the effect of sucking the earth's radiation belts down to where they contact the atmosphere, with the same results as above. This is called a geomagnetic storm. Monitoring of the sun can give as much as a days warning of the arrival of charged particles from a solar storm, but geomagnetic activity does not telegraph its' punch in this way. This means that the orbit of a satellite suffering high levels of air drag is very uncertain when propagated into the future, since the future state of solar and geomagnetic activity cannot be predicted.

4.8 Radiation Pressure

In very high accuracy work, the pressure of solar radiation on a satellite must sometimes be considered. The acceleration of the satellite depends on the area of the satellite A and its distance from the sun R_S as

$$\mathbf{a}_l = \frac{E}{c}\frac{Ag}{m}\frac{\mathbf{R}_S}{R_S^3} \tag{4.58}$$

Of course, the intensity of sunlight depends on the distance from the sun as an inverse square law, and if the satellite reflects the light diffusely, the acceleration will be directed away from the sun. The quantity E is called the solar constant, and is the amount of energy flowing through a 1 cm surface at 1 A.U. from the sun. As energy E and momentum p are related by $E = pc$ for light, where c is the speed of light, E/c gives the momentum flow per unit area per unit time. The speed of light, c, is approximately 2.998×10^{10} cm/sec, which is a very large number. Hence, E/c is a small number.

The quantity g is the reflectivity factor, and plays a role similar to C_D in the air drag law: it includes the possibility of different reflection directions

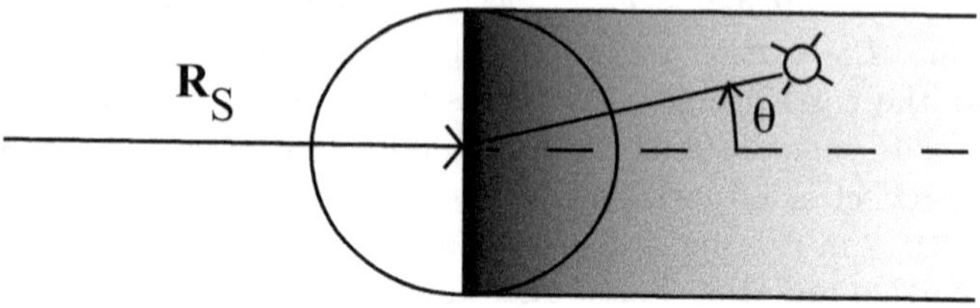

Figure 4.7: Earth Shadow Geometry

due to the shape of the object. A planar object oriented broadside to the sun would have $g = 1$ if it is a perfect absorber of sunlight, since all of the incident momentum would be soaked up by the object. Even better, however, is a perfect reflector, since a mirror will reverse the direction of the momentum vector for a reflected photon. A perfect planar reflector would have $g = 2$, or just twice as large.

The susceptibility of a satellite to radiation pressure is governed by the factor Ag/m, which depends on the area to mass ratio just like the air drag factor B^*. The Echo and Pageos balloon satellites were heavily perturbed by sunlight, since they were about 100 feet in diameter (large A), made of 1/2 mil plastic (small m), and coated with aluminum ($g \approx 2$).

One additional complication of the radiation pressure force is that it vanishes when a satellite slips into the earth's shadow. As shown in Figure 4.7, for low earth satellites we must have

$$\cos\theta = \frac{\mathbf{R_S} \cdot \mathbf{r}}{R_S\, r} > 0 \tag{4.59}$$

for the satellite to be on the "outward" side of the earth, and

$$r\cos\theta < R_\oplus \tag{4.60}$$

for the satellite to be in shadow. For very high orbits, the shadow conditions are more complicated, since the shadow is really a long, narrow cone.

For objects in solar orbit, there is a simplification which occurs. Since the pressure of sunlight obeys a repulsive inverse square law, its only effect on an object orbiting the sun is to make the sun seem less massive. If we write the equations of motion for an object orbiting the sun, including the two body force and light pressure, we obtain

$$\ddot{\mathbf{R}}_S = -\left(\mu - \frac{EAg}{cm}\right)\frac{\mathbf{R_S}}{R_S^3} \tag{4.61}$$

For a typical interplanetary spacecraft, this appears to change the mass of the sun slightly. All the usual two – body dynamics still apply, however. But, if $EAg/cm > \mu$, then the sun appears to have a negative mass. In this case, the only permitted conic section orbit is the branch of the hyperbola which is forbidden in normal two – body gravitational motion. Then, all orbits become hyperbolic escape orbits, and the spacecraft becomes an interstellar probe.

Still more esoteric possibilities exist for light to perturb orbits. In the Yarkovsky effect, a satellite warms on the sunlit side, and cools on the night-time side. As it rotates, the point of maximum re-radiation of infrared light occurs in the "late afternoon". This creates an acceleration on the spacecraft that is not simply directed away from the sun. Comets experience a similar effect as their ices sublime, and this produces a reaction force on the comet nucleus. Also, the geometry of the vehicle can cause a similar effect with specular reflection in preferred directions. It has been discovered that this asymmetric reflection effect can also alter the spin state of small asteroids, even to spinning them up to the point that they fission and produce a satellite. Finally, the Poynting-Robertson effect is a relativistic effect whereby a very small object, like a dust grain, is heated uniformly by the sun, and radiates this energy back into space isotropically, *in its own rest frame.* However, in the inertial frame the radiative energy is slightly concentrated in the forward direction, causing small dust grains to slowly spiral into the sun.

4.9 References

Abramowitz and Segun, "Handbook of Mathematical Functions" is an invaluable reference on Legendre polynomials and many other such functions. It exists in both Dover paperback and Government Printing Office versions, the latter in hardcover. Legendre polynomials are found in Abramowitz and Segun, chapter 8.

The current best publically available Earth gravity model is the Earth Gravity Model 96, or "EGM96" for short. It is, at the time of this writing, conveniently found on the world wide web at

http://cddisa.gsfc.nasa.gov/926/egm96/egm96.html

which makes it possible to download the gravity coefficients in machine readable form. More recent models may be available after the time of this writing, of course

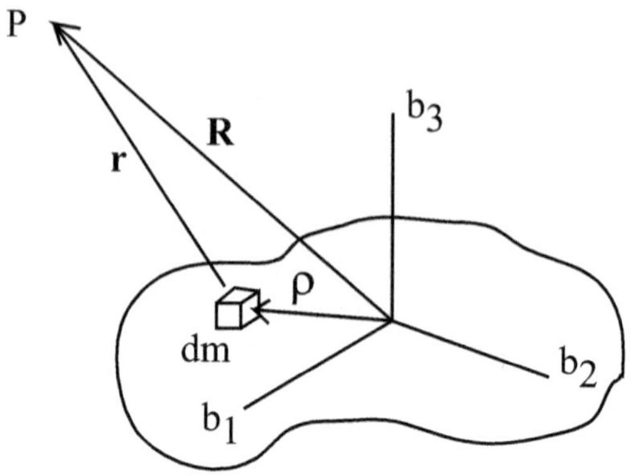

Figure 4.8: Mass Integral Potential Expansion

4.10 Problems

Problem 1. If the independent variable is changed from θ to $x = \cos\theta$, show that Legendre's equation takes the form

$$\frac{d}{dx}\left((1-x^2)\frac{d}{dx}\Theta\right) + \left(n(n+1) - \frac{m^2}{1-x^2}\right)\Theta = 0$$

Also, transform the boundary condition. Verify that the first three Legendre polynomials P_0^0, P_1^0, and P_2^0 are solutions with the correct boundary conditions.

Problem 2. An alternative to expanding the geopotential in spherical harmonics is the expansion in mass integrals. This is especially useful for rotational dynamics of orbiting bodies, and for expressing the potential field of objects (eg, asteroids) which are too small to force themselves to be nearly spherical. If the **b** frame in Figure 4.8 is the principal axis frame of the mass M, (not spherical!), argue that the potential per unit mass at point p is given by

$$V = -G\int \frac{dm}{r} = -G\int \frac{dm}{(R^2 - 2\mathbf{R}\cdot\vec{\rho} + \rho^2)^{1/2}}$$

Write

$$\mathbf{R} = R(l\mathbf{b}_1 + m\mathbf{b}_2 + n\mathbf{b}_3)$$

where l, m, n are the direction cosines of $\mathbf{R}$, and

$$\vec{\rho} = x\mathbf{b}_1 + y\mathbf{b}_2 + z\mathbf{b}_3$$

Expand the denominator of V to the *second* order using the binomial theorem, and show that the potential function becomes

$$V \approx -\frac{GM}{R} - \frac{GM}{4R^3}\{(3l^2-1)(B+C-A) + (3m^2-1)(A+C-B)$$
$$+(3n^2-1)(A+B-C)\} + \mathcal{O}(R^{-5})$$

where A, B, C are the principal moments of inertia of the mass M.

Problem 3. Derive the Hamiltonian for satellite motion about a non – spherical planet using polar coordinates r, θ and ϕ_{long} and employing the full geopotential expansion. That is, set up the problem in a reference frame rotating with the earth's angular velocity $\omega_{\oplus}$. Show that this form of the problem possesses an integral of the motion.

Chapter 5

General Perturbations

5.1 Introduction

Most systems encountered in orbital mechanics are nearly perfect two body problems, with only small perturbations from two body motion. There are obvious exceptions to this, as in the case of an interplanetary spacecraft. If we leave the vicinity of the earth on a nearly hyperbolic trajectory (with respect to the earth), follow a nearly elliptical transfer trajectory to another planet (with respect to the sun), and then approach the other planet on a nearly hyperbolic flyby trajectory (with respect to the planet), there is no way to model the entire orbit as a two body orbit with respect to one object plus small perturbations.

However, when the perturbations are small we can make some approximations. *The fundamental assumption of perturbation theory is that the perturbations remain small. When this is not true the perturbation results are invalid.* If the perturbations remain small, then the solution for the unperturbed motion can be used to evaluate the right hand sides of the equations of motion. With some additional approximations, usually involving series expansions in the (assumed) small eccentricity e, we may obtain forms which can be integrated to give approximate solutions.

In this chapter we will discuss the general approach to perturbation theory, both from the classical and canonical perspective. Several examples will be considered: we will complete the first order solution to the simple pendulum considered as an "osculating harmonic oscillator". The first order solution to the "main problem" of artificial satellite theory: the problem of a satellite perturbed by the J_2 oblateness of the earth, will be a major concern in this chapter, as will third body perturbations on the orbit. We will also briefly look at third body perturbations on the *rotation* of a satellite, and finish the chapter with a first look at one systematic approach to higher order perturbations.

5.2 Canonical Perturbation Theory

Perturbation theory is most elegantly posed in canonical variables, using canonical transformations. Suppose we have the Hamiltonian of a system

$$H(q_i, p_i, t) = H_o(q_i, p_i, t) + \epsilon H_1(q_i, p_i, t) \tag{5.1}$$

in terms of the variables q_i, p_i. Suppose the first term represents the "dominant" part of the Hamiltonian, and that it can be solved in closed form. For orbital mechanics, this will almost always be the two body part of the system. The second term contains a small parameter ϵ, which makes this part less important. This might be the mass of a small "third" gravitating body, a small air drag parameter, a small coordinate or momentum, or the inverse of a large distance. Also, suppose that we have carried out the Hamilton – Jacobi procedure on the dominant portion of the Hamiltonian, and have found the new variables Q_i, P_i which cause K_o to vanish. Of course, if the system really just consisted of the dominant terms H_o, these new coordinates and momenta would all be constants of the motion. Usually, H_o is called the zero order part of the system, while H_1 is termed the first order part of the Hamiltonian.

Now, if we apply this same transformation to the system (5.1), including the perturbation term, the result will be

$$K(Q_i, P_i, t) = \epsilon K_1(Q_i, P_i, t) \tag{5.2}$$

The dominant portion will still vanish under the transformation, and all that remains is the perturbation term $\epsilon K_1(Q_i, P_i, t)$. In fact, it is easy to see that K_1 is simply H_1 transformed into the new variables. The new equations of motion are

$$\dot{Q}_i = \epsilon \frac{\partial K_1}{\partial P_i}, \quad \dot{P}_i = -\epsilon \frac{\partial K_1}{\partial Q_i} \tag{5.3}$$

These express the fact that the new coordinates and momenta change only because of the perturbations present in the full system (5.1). Notice, however, that any terms from the zero order part of the Hamiltonian have vanished. These are still fully valid equations of motion for the complete system (5.1). We have made no approximations at all....just a change of variables from the "natural" variables q_i, p_i to the better, but 'nonobvious', variables Q_i, P_i. So equations (5.3) are still likely to be highly coupled, highly nonlinear ordinary differential equations. Although the new variables change only slowly (that is what the factor of ϵ out front says), these equations are probably no more solvable in closed form than the original full system (5.1).

However, the unperturbed system was solved exactly by this coordinate transformation, with the result that $Q_i = Q_{io}$, all constant, and $P_i = P_{io}$, also all constant. The new equations of motion also confirm that although Q_i, P_i do change, they do so slowly. If the new variables are approximately constant, then we make only a small error if we substitute this assumption into the equations of motion, obtaining

$$\begin{aligned} \dot{Q}_i &\approx \epsilon \frac{\partial}{\partial P_{io}} K_1(Q_{io}, P_{io}, t) \\ \dot{P}_i &\approx -\epsilon \frac{\partial}{\partial Q_{io}} K_1(Q_{io}, P_{io}, t) \end{aligned} \tag{5.4}$$

Notice that it is right here that we have applied the fundamental assumption of perturbation theory: the hope that the perturbations remain small. If they do *not* remain small, equations (5.4) are nonsense, and we would need to revert to the study of equations (5.3). However, a marvelous change has occurred with this substitution. The equations (5.3) are fully coupled, nonlinear ordinary differential equations, which we can write symbolically as

$$\dot{\mathbf{x}} = \mathbf{f}(\mathbf{x}, t) \tag{5.5}$$

On the other hand, the Q_{io} and P_{io} on the right sides of (5.3) are constants, so these equations can be written symbolically as

$$\dot{\mathbf{x}} = \mathbf{f}(c_i, t) \tag{5.6}$$

where the c_i are constants. The rights sides of the equations of motion are no longer coupled. In fact, they are functions of time alone.

There is no general procedure for solving the systems (5.3) or (5.5), but the system (5.6) can be solved by separation of variables and integration:

$$\mathbf{x}(t) = \mathbf{x}(t_o) + \int_{t_o}^{t} \mathbf{f}(c_i, t) dt \tag{5.7}$$

Returning to the system (5.3), the solution can be written

$$\begin{aligned} Q_i &\approx Q_{io} + \epsilon \int_{t_o}^{t} \frac{\partial}{\partial P_{io}} K_1(Q_{io}, P_{io}, t) dt \\ P_i &\approx P_{io} - \epsilon \int_{t_o}^{t} \frac{\partial}{\partial Q_{io}} K_1(Q_{io}, P_{io}, t) dt \end{aligned} \tag{5.8}$$

Now, there is no guarantee that we will be clever enough to actually carry out these integrals in closed form. Even owning a large integral table may not be sufficient. So, in order to perform the integrals we may be forced to resort to any legal trick to extract an approximate closed form result. Series expansions of the integrands is one important option, especially if some of the variables Q_i, P_i are themselves 'small'. The usual expansion variables in orbital problems are the eccentricity e, and perhaps the ratio of the semimajor axis a to the large distance to a 'third' body.

If we can perform the integrals above, we now have in our possession the "first order perturbation solution" to the problem. The solution to (5.8) will furnish values of the "osculating elements", and these can be substituted into the original coordinate transformation to obtain values of the physical coordinates q_i, p_i correct to the first order in ϵ. Now, our original assumption that Q_{io}, P_{io} were constant may return to haunt us. The approximate solution (5.8) at best shows that Q_i, P_i do change with an amplitude proportional to ϵ. We could then return to the system (5.3) and improve our constant variable assumption by substituting the entire first order solution into the right hand sides. This will lead to the second order perturbation solution to the problem, since terms proportional to ϵ^2 will now appear in the result. However, the results (5.8) could also contain infinities, or terms so large as to violate our original assumption of small perturbations. This would invalidate our substitution of constant Q_{io}, P_{io} on the right sides of (5.3), and destroy our confidence in the results.

Now, this has been pretty abstract, and it does not really need to be canonical. The same sequence of operations can be carried out with the Lagrange Planetary equations using the classical orbital elements. The Delaunay elements are often used in canonical perturbation theory, in which case the zero order Hamiltonian doesn't quite vanish. The best way to become familiar with the technique is to see (and do !) examples. So, let us begin.

5.3 The Simple Pendulum, II

In section 3.7 we have already carried out the first steps in attacking the simple pendulum as a perturbation problem. Although the simple pendulum can be solved completely in closed form (for fans of elliptic integral functions), we will find it instructive to tackle it using perturbation theory techniques. In the earlier section we introduced a new coordinate Q and momentum P, such that the transformation back to the physical angle θ was

$$\theta = \sqrt{\frac{2P}{\sqrt{gl^3}}} \sin Q \tag{5.9}$$

and, after expanding the $\cos\theta$ potential term and transforming to the new variables, the Hamiltonian became

$$K = \sqrt{\frac{g}{l}} P - gl \sum_{n=2}^{\infty} \frac{(-1)^n}{(2n)!} \left(\frac{2P}{\sqrt{gl^3}} \right)^n \sin^{2n} Q \tag{5.10}$$

If we restrict ourselves to the harmonic oscillator part of the Hamiltonian, the new momentum P is constant, say P_o, while the new coordinate Q is a linear function of time

$$Q = \sqrt{\frac{g}{l}} t + Q_o = \omega t + Q_o \tag{5.11}$$

where ω is the harmonic oscillator frequency. Note that for small amplitudes, the new momentum P must be small. This serves the role of the small quantity ϵ in the last section.

Before proceeding with the perturbation solution to this system, it pays to think ahead to the problem of performing the definite integrals we will find. Part of this work has already been done when we expanded the $\cos\theta$ term in the original Hamiltonian. Let us group the messy coefficients in the expansion above as

$$c_n = gl\frac{(-1)^n}{(2n)!}\left(\frac{2}{\sqrt{gl^3}}\right)^n \tag{5.12}$$

Also, it is easier to integrate sines and cosines of multiple angles, rather than integrating powers of sines and cosines. Replace the sine term with its expansion in terms of multiple angles.

$$\sin^{2n} Q = \sum_{j=0}^{2n} \beta_{nj} \cos jQ \tag{5.13}$$

The β_{nj} coefficients are simply numbers: since

$$\sin^2 Q = 1/2 - 1/2\cos 2Q$$

we have $\beta_{10} = 1/2, \beta_{11} = 0, \beta_{12} = -1/2$, for example. Also, as another example,

$$\sin^4 Q = 3/8 - 1/2\cos 2Q + 1/8\cos 4Q$$

So, the three nonzero coefficients are $\beta_{20} = 3/8, \beta_{22} = -1/2$, and $\beta_{24} = 1/8$. The Hamiltonian then becomes

$$K = \omega P - \sum_{n=2}^{\infty}\sum_{j=0}^{2n} c_n \beta_{nj} P^n \cos(jQ) \tag{5.14}$$

The next step is to find the equations of motion. These are simply the usual Hamilton's equations:

$$\dot{Q} = \omega - \sum_{n=2}^{\infty}\sum_{j=0}^{2n} c_n \beta_{nj} n P^{n-1} \cos(jQ) \tag{5.15}$$

$$\dot{P} = -\sum_{n=2}^{\infty}\sum_{j=0}^{2n} c_n \beta_{nj} P^n j \sin(jQ) \tag{5.16}$$

and are much more messy than the relatively simple

$$\dot{\theta} = p/l^2, \quad \dot{p} = -gl\sin\theta \tag{5.17}$$

which we would find in the original variables. However, equations (5.15) and (5.16) are still exact, since the series expansions used are convergent.

If we now substitute the harmonic oscillator solution in the right sides of (5.15) and (5.16), we have

$$\dot{Q} = \omega - \sum_{n=2}^{\infty}\sum_{j=0}^{2n} c_n \beta_{nj} n P_o^{n-1} \cos j(\omega t + Q_o) \quad (5.18)$$

$$\dot{P} = -\sum_{n=2}^{\infty}\sum_{j=0}^{2n} c_n \beta_{nj} P_o^n j \sin j(\omega t + Q_o) \quad (5.19)$$

which is not getting any simpler. They are also invalid if the departures from 'harmonic oscillator' motion do not remain small. But the advertised decoupling of the equations of motion has occurred, and the right sides of (5.18) and (5.19) are functions of time alone.

The solution is now obtained by calculating the definite integrals

$$P = P_o + \int_0^t \dot{P} dt, \quad Q = Q_o + \int_0^t \dot{Q} dt \quad (5.20)$$

However, it pays to be slightly careful. The $j = 0$ terms in the P equation are not really present, so we can begin this summation at $j = 1$. Also, we might be tempted to integrate the $j = 0$ cosine terms in the Q equation as

$$\int \cos 0Q dt = \frac{1}{0} \sin 0Q \quad (5.21)$$

which is also not correct. The time dependence vanishes for these $j = 0$ terms, and they are constant. Terms of this type are called *secular terms,* since they will integrate to linear functions of time.

The first order solution is easily obtained as

$$\begin{aligned} Q &= \omega t + Q_o - \left(\sum_{n=2}^{\infty} n c_n \beta_{n0} P_o^{n-1}\right) t \\ &\quad - \sum_{n=2}^{\infty}\sum_{j=1}^{2n} \frac{c_n \beta_{nj} n}{j\omega} P_o^{n-1} \sin j(\omega t + Q_o)\Big|_0^t \end{aligned} \quad (5.22)$$

$$P = P_o + \sum_{n=2}^{\infty}\sum_{j=1}^{2n} \frac{c_n \beta_{nj}}{\omega} P_o^n \cos j(\omega t + Q_o)\Big|_0^t \quad (5.23)$$

Notice that the first one or two terms of each solution simply return to us the harmonic oscillator part of the solution, also called the zero order solution.

The secular terms appear only in the Q solution, where they have the effect of simply changing the period of the pendulum's oscillation. The period is now given by

$$T = \frac{2\pi}{\omega - \sum_{n=2}^{\infty} n c_n \beta_{n0} P_o^{n-1}} \quad (5.24)$$

which depends on the amplitude P. Secular terms can be dangerous, since they can destroy the "small perturbation" assumption we have made. However, they can be tolerated in this case, since the physical coordinate does not deviate greatly from the assumed harmonic oscillator solution even with the presence of the secular terms. Luckily, they do not appear in the momentum solution, where they would predict an ever growing or decreasing amplitude, something which we would not expect a pendulum to do without friction.

We are now free to repeat the process. The first order solution can be substituted into the right hand sides of the equations of motion (5.15) and (5.16), further expansions performed, and the result integrated to give the second order perturbation solution. A few moments of contemplating this process will reveal why so many perturbation solutions are truncated at the first order.

Also, this process can never yield the full solution to the simple pendulum. The pendulum is a system which oscillates for amplitudes $\theta < \pi$, and goes into a state of rotation if the amplitude exceeds this value. However, we have attacked this problem as an osculating harmonic oscillator problem. The form of the coordinate transformation (5.9) absolutely forbids the appearance of rotational behavior. At best, we can hope that this process will converge up to the limit where the real pendulum transitions to rotation. This perturbation method cannot enter the rotation area of the phase space. However, another perturbation attack can be carried out, starting from the assumption that the pendulum is in a very high, nearly constant speed state of rotation. This method would probably converge down to the transition to oscillation.

5.4 Expansions in Elliptic Motion

Just as we found it necessary in the last section to expand the equations of motion in order to obtain an integrable form, it is usually necessary to expand the perturbing force in order to obtain integrable expressions for orbit perturbation problems. Given the choice, it is generally better to expand the potential energy (or its negative, the disturbing function) rather than to expand six separate equations of motion. Here, we will take a somewhat heuristic approach, since our goal will only be to obtain results correct through the second order in the eccentricity.

Let us begin with the reputably unsolvable Kepler's equation. Rewrite it as

$$E = M + e \sin E \tag{5.25}$$

and notice that if $e = 0$, we have the solution $E = M$. For small eccentricity, the second term above is not large, so we only commit a small error if we substitute this into the left side of (5.25), to obtain

$$E \approx M + e \sin M \tag{5.26}$$

This result is now approximately correct through the first order in the eccentricity. Substitute this result into (5.25) to find

$$E \approx M + e\sin(M + e\sin M) \tag{5.27}$$

which contains the very unpretty term $e\sin(M + e\sin M)$. However, the term $e\sin M$ is still assumed to be small, so we can apply the sum law for sines to get

$$\sin(M + e\sin M) = \sin M\cos(e\sin M) + \cos M\sin(e\sin M) \tag{5.28}$$

and then expand the sine and cosine of the small angle as

$$\cos(e\sin M) \approx 1, \quad \sin(e\sin M) \approx e\sin M \tag{5.29}$$

We only need these results correct to the first order in e, since the offending term already contains the first power of the eccentricity. Finally, grouping terms we find the approximate solution to Kepler's equation as

$$E \approx M + e\sin M + \frac{1}{2}e^2\sin 2M + \mathcal{O}(e^3) \tag{5.30}$$

correct through the second order. The notation $\mathcal{O}(e^3)$ indicates that terms of the third order and higher have been truncated. By continuing this process, the solution can be constructed to any order.

Now, take the law for converting from eccentric anomaly E to true anomaly ν. Expand the denominator with the binomial theorem assuming small eccentricity, to find

$$\begin{aligned} \cos\nu &= \frac{\cos E - e}{1 - e\cos E} = (\cos E - e)(1 - e\cos E)^{-1} \\ &\approx (\cos E - e)(1 + e\cos E + e^2\cos^2 E + \ldots) \end{aligned} \tag{5.31}$$

Multiply out the result, finding

$$\cos\nu \approx \cos E - e\sin^2 E + e^2(\cos^3 E - \cos E) + \ldots \tag{5.32}$$

After eliminating powers of sines and cosines in favor of their multiple angle equivalents, this becomes

$$\cos\nu \approx \left(1 - \frac{1}{4}e^2\right)\cos E + \frac{1}{2}e\cos 2E + \frac{1}{4}e^2\cos 3E + \mathcal{O}(e^3) \tag{5.33}$$

It is somewhat more difficult to obtain the direct conversion between the true and mean anomalies

$$\nu \approx M + 2e\sin M + \frac{5}{4}e^2\sin 2M + \mathcal{O}(e^3) \tag{5.34}$$

although this is often the more useful result.

As a final result, we will need the expansion of the radius in terms of the mean anomaly. Begin with the expression for r in terms of the eccentric anomaly

$$r = a(1 - e\cos E) \tag{5.35}$$

and substitute our approximate solution for E

$$r \approx a(1 - e\cos(M + e\sin M)) \tag{5.36}$$

which is enough to yield a result correct to second order. The cosine can be broken apart with the multiple angle formula

$$\cos(M + e\sin M) = \cos M\cos(e\sin M) - \sin M\sin(e\sin M) \tag{5.37}$$

and the sine and cosine of the small angles expanded to the second order in e, obtaining

$$r \approx a\left(1 + \frac{1}{2}e^2 - e\cos M - \frac{1}{2}e^2\cos 2M + \mathcal{O}(e^3)\right) \tag{5.38}$$

The topic of expansions in elliptic motion normally has a chapter to itself in most advanced celestial mechanics texts. It is the technique which made celestial mechanics possible in the three centuries before the invention of the computer, and much effort has been devoted to it. The solution to Kepler's equation, for example, can be expressed to any order in e through use of the Bessel functions. However, much old–time celestial mechanics work involved the drudgery of the hand algebra of truncated infinite series. We will continue with this approach, to give you a feel for the accomplishments of the heroic age of this discipline, but modern series expansion work would be done with a computer program designed to perform literal algebraic manipulation of series expansions.

5.5 The J_2 Disturbing Function

For a satellite outside the earth's atmosphere, the dominant perturbation is the earth's oblateness, or the J_2 term in the geopotential. This term is at least three orders of magnitude larger than any other term in the geopotential. In fact, it is so important that the problem of solving for the J_2 perturbation effects has earned the name of the "main problem of artificial satellite theory". Many solutions to this problem are available, in both the classical and canonical elements; including advanced versions which perform the series expansions in near closed form with "Kozai coefficients", and even a few solutions which do not need series expansions at all. However, in this and following sections we will take the most obvious approach of using the classical orbital elements and simple series expansion techniques.

The first problem is to take the J_2 term of the geopotential expansion

$$R_2 = -\frac{\mu R_\oplus^2 J_2}{2r^3}\left(3\cos^2\theta - 1\right) \tag{5.39}$$

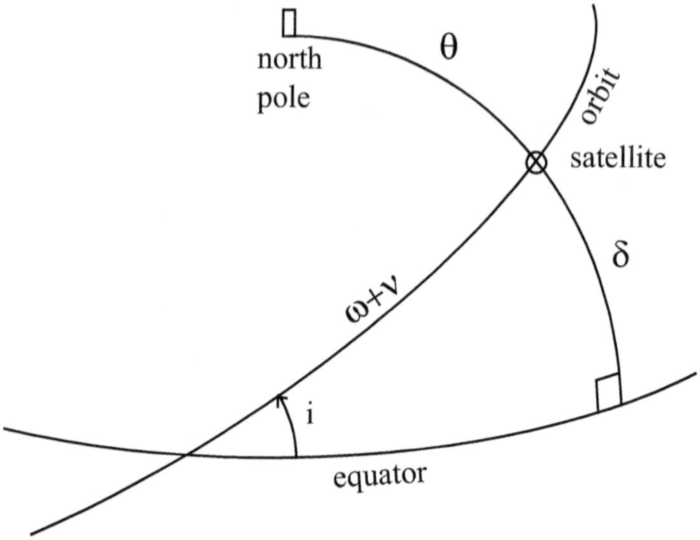

Figure 5.1: Elimination of the Colatitude

and convert it into the classical orbital elements. This must be done before we can calculate the partial derivatives required for the Lagrange Planetary equations. Also, we will almost certainly need to perform some series expansions to obtain a form of the perturbation equations that can be integrated. This is most easily done by expanding the disturbing function itself, rather than performing separate expansions on each equation of motion.

As a first step we will begin by disposing of the colatitude θ in (5.39). Applying the sine law of spherical trigonometry to the spherical triangle shown in Figure 5.1, we have

$$\cos\theta = \sin\delta = \frac{\sin i \sin(\omega + \nu)}{\sin 90^o} \tag{5.40}$$

The disturbing function now becomes

$$R_2 = -\frac{\mu R_\oplus^2 J_2}{2r^3}\left(3\sin^2 i \sin^2(\nu + \omega) - 1\right) \tag{5.41}$$

This introduces the elements i and ω, and the remaining replacements involve converting $r(t), \nu(t)$ to the classical elements.

Let us tackle the radial term first. Recalling the radial expansion from the last section, we have

$$r^{-3} = a^{-3}\left(1 - e\cos M + \frac{1}{2}e^2 - \frac{1}{2}e^2\cos 2M\right)^{-3} \tag{5.42}$$

Expanding the bracketed term via the binomial theorem, and retaining terms through the second order gives

$$\begin{aligned} r^{-3} &= a^{-3}\left(1 - 3(-e\cos M + \frac{1}{2}e^2 - \frac{1}{2}e^2\cos 2M)\right. \\ &+ \left.\frac{(-3)(-4)}{2!}e^2\cos^2 M\right) + \mathcal{O}(e^3) \end{aligned} \tag{5.43}$$

Finally, replacing the powers of trigonometric functions with their multiple angle equivalents gives

$$r^{-3} = a^{-3}\left(1 + 3e\cos M + \frac{3}{2}e^2 + \frac{9}{2}e^2\cos 2M\right) \tag{5.44}$$

The true anomaly term requires more labor. Begin by replacing $\sin^2(\nu+\omega) = 1/2 - 1/2\cos 2(\nu+\omega)$, and then insert the expansion for the true anomaly, obtaining

$$\cos(2\nu + 2\omega) \approx \cos(2M + 2\omega + 4e\sin M + \frac{5}{2}e^2\sin 2M) \tag{5.45}$$

The angle $2M + 2\omega$ is large, while the remainder of the argument of the cosine is small, at worst of order e. Using the trigonometric identity for the cosine of a sum, and then replacing the sine and cosine with their small angle approximations to second order, we find

$$\begin{aligned} \cos(2\nu + 2\omega) &\approx \cos(2M + 2\omega)\cos(4e\sin M + \frac{5}{2}e^2\sin 2M) \\ &- \sin(2M + 2\omega)\sin(4e\sin M + \frac{5}{2}e^2\sin 2M) \\ &\approx \cos(2M + 2\omega)(1 - 8e^2\sin^2 M) \\ &- \sin(2M + 2\omega)(4e\sin M + \frac{5}{2}e^2\sin 2M) \end{aligned} \tag{5.46}$$

Finally, the products of sines and cosines can be replaced with multiple angle equivalents, giving

$$\begin{aligned} \cos(2\nu + 2\omega) &\approx \cos(2M + 2\omega) - 2e\left[\cos(M + 2\omega)\right. \\ &- \left.\cos(3M + 2\omega)\right] - 4e^2\cos(2M + 2\omega) \\ &+ \frac{3}{4}e^2\cos 2\omega + \frac{13}{4}e^2\cos(4M + 2\omega) \end{aligned} \tag{5.47}$$

The final step is to insert these results into the disturbing function, and multiply the two series together. This is laborious, but the work is considerably shortened since we are only retaining terms through the second order. After again replacing trigonometric products with their multiple angle equivalents, the final result is

$$\begin{aligned}
R_2 &= -\frac{\mu R_\oplus^2 J_2}{2a^3}\left(\frac{3}{2}\sin^2 i - 1 - \frac{3}{2}e^2 + \frac{9}{4}e^2\sin^2 i\right. \\
&- \frac{3}{2}\sin^2 i\cos(2M+2\omega) + \frac{3}{4}e\sin^2 i\cos(M+2\omega) \\
&- \frac{21}{4}e\sin^2 i\cos(3M+2\omega) + \frac{9}{2}e\sin^2 i\cos(M) \\
&- 3e\cos M - \frac{51}{4}e^2\sin^2 i\cos(4M+2\omega) \\
&+ \frac{15}{4}e^2\sin^2 i\cos(2M+2\omega) - \frac{9}{2}e^2\cos(2M) \\
&+ \left.\frac{27}{4}e^2\sin^2 i\cos(2M)\right) + \mathcal{O}(e^3)
\end{aligned} \tag{5.48}$$

which is correct through the second order in the eccentricity. This will hopefully be adequate for low earth satellites, since proximity to the earth forces low eccentricity. The portion of (5.48) independent of the mean anomaly M is

$$R_{2,sec} = -\frac{\mu R_\oplus^2 J_2}{2a^3}\left(\frac{3}{2}\sin^2 i - 1\right)\left(1+\frac{3}{2}e^2\right) \tag{5.49}$$

This is the portion of the disturbing function which will yield secular terms in the solution.

5.6 The Secular J_2 Terms

Only one of the classical elements we are using varies with time, and that is the mean anomaly. Since the next step of the perturbation solution will be to evaluate the equations of motion and substitute the two body solution, only terms containing the mean anomaly M will have explicit time dependence. These terms are called periodic terms, since the mean anomaly dependence is always contained within a sine or cosine, and the result of the final time integration will be a cosine or a sine. So, the periodic terms contribute periodic oscillations in each of the elements which remain small. However, terms in the expansion of the disturbing function which do not contain M are secular terms, since they do not lead to small changes in the orbit. Rather, if the secular term survives the process of taking the partial derivative in the Lagrange Planetary equations, we must then assume that it is constant, and we will be forced to integrate it into a term proportional to time in the final result. This could lead to perturbations which do not remain small, violating the fundamental assumption of perturbation theory.

The secular terms from the J_2 disturbing function are

$$R_{2,sec} = -\frac{\mu R_\oplus^2 J_2}{2a^3}\left(1+\frac{3}{2}e^2\right)\left(\frac{3}{2}\sin^2 i - 1\right) \tag{5.50}$$

More advanced expansion techniques will show that this is just the first four terms of

$$R_{2,sec} = -\frac{\mu R_\oplus^2 J_2}{2a^3(1-e^2)^{3/2}}\left(\frac{3}{2}\sin^2 i - 1\right) \tag{5.51}$$

which is correct to any order of e and i. To find this, we can either collapse the first two terms of the binomial expansion of $(1-e^2)^{-3/2}$, or notice that the secular part is free of the mean anomaly, so it must be the constant term in the Fourier series of R_2:

$$R_{2,sec} = \frac{1}{2\pi}\int_0^{2\pi} R_2(M)dM \tag{5.52}$$

In the form (5.41), replace $\sin^2(\omega+\nu) = (1-\cos 2(\omega+\nu))/2$ to find

$$R_2 = -\frac{\mu J_2 R_\oplus^2}{2r^3}\left(\left[\frac{3}{2}\sin^2 i - 1\right] - \frac{3}{2}\sin^2 i \cos 2(\omega+\nu)\right) \tag{5.53}$$

The first averaging integral

$$I_1 = \frac{1}{2\pi}\int_0^{2\pi}\frac{1}{r^3}dM \tag{5.54}$$

can be converted to the eccentric anomaly using $r = a(1-e\cos E)$, and since $M = E - e\sin E$ we have $dM = (1-e\cos E)dE$. This gives

$$I_1 = \frac{a^{-3}}{2\pi}\int_0^{2\pi}\frac{dE}{(1-e\cos E)^2} = -\frac{1}{a^3(1-e^2)^{3/2}} \tag{5.55}$$

The last step above can be carried out by any modern mathematics package, or by complex contour integration and the Cauchy integral theorem. The second integral

$$I_2 = \frac{1}{2\pi}\int_0^{2\pi}\frac{\cos 2(\omega+\nu)}{r^3}dM = 0 \tag{5.56}$$

is zero by symmetry, and somewhat more laboriously by conversion to the true anomaly as integration variable. This gives (5.51) as a final result.

Note that the secular part of the disturbing function contains only the elements a, e, and i. Returning to the Lagrange Planetary equations in section 3.8, we see that the only classical elements which have secular terms in their perturbation solutions are the node Ω, the argument of perigee ω, and the mean anomaly M.

For the moment, let us proceed as if we were not worried that the secular terms will invalidate the final result. Substituting (5.51) into the Lagrange Planetary equations, we find expressions for the secular rates in ω, Ω, and M as

$$\dot{\omega} = -\frac{3nJ_2R_\oplus^2}{2a^2(1-e^2)^2}\left(\frac{5}{2}\sin^2 i - 2\right) \tag{5.57}$$

$$\dot{\Omega} = -\frac{3nJ_2R_\oplus^2}{2a^2(1-e^2)^2}\cos i \tag{5.58}$$

$$\dot{M}_o = -\frac{3nJ_2R_\oplus^2}{2a^2(1-e^2)^{3/2}}\left(\frac{3}{2}\sin^2 i - 1\right) \tag{5.59}$$

The next step is to declare that the expressions on the right sides of the above have their two – body values. Since these are now all constants, we are forced to integrate them with respect to time in the form

$$\omega = \omega_o + \dot{\omega}t \tag{5.60}$$

$$\Omega = \Omega_o + \dot{\Omega}t \tag{5.61}$$

$$M = M_o + (n + \dot{M}_o)t \tag{5.62}$$

and the perturbations grow without limit in time. However, as we saw in the case of the pendulum, this is not necessarily a problem.

In particular, a secular term in the mean anomaly M simply has the effect of modifying the two – body period. The time it takes the satellite to move from perigee to perigee (the anomalistic period, or the period of the anomalies) is now given by

$$T_p = \frac{2\pi}{n + \dot{M}_o} \tag{5.63}$$

This is also referred to as the Kozai period of the orbit, and the modified mean motion

$$n_K = n + \dot{M}_o \tag{5.64}$$

is termed the Kozai mean motion. So, the secular term in the mean anomaly simply has the effect of changing the period of the orbit.

The secular term in the node implies that the orbit plane regresses. This is also a familiar effect from observing the motion of a child's top. When a spinning gyroscope is subject to a torque trying to topple it over, it responds by precessing. As shown in Figure 5.2, the case of satellite motion is analogous. The spinning rotor is, in some averaged sense, the satellite in its orbit. The external torque is supplied by the additional mass in the earth's equatorial bulge. The additional acceleration due to the mass in the equatorial bulge contributes inward accelerations when crossing the equator, but at the points on the orbit furthest from the equator the additional mass attempts to pull the satellite back towards the equator. Averaged around the orbit, this gives a net torque $\mathbf{M}$ which would like to rotate the orbit plane into the plane of the equator. However, the result is precession of the orbit plane, since the net torque $\mathbf{M}$ is the inertial rate of change of the orbital angular momentum $\mathbf{h}$.

The final secular effect is that the orbit itself rotates within the orbital plane. This is also a gyroscopic effect, although it is not as familiar as is precession. None of these effects is disastrous to the final perturbation solution. This is in contrast to what would have happened if any of the other three elements had had a secular term. A secular term in the semimajor axis a would lead to

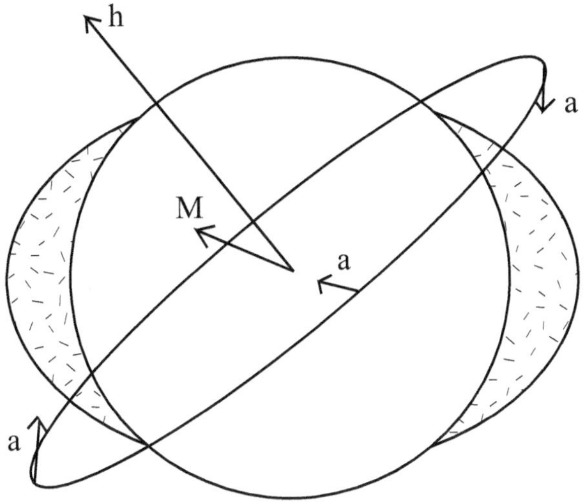

Figure 5.2: Regression of the Nodes

unlimited expansion or contraction of the orbit, an effect which is not expected in a system which still conserves energy. A secular term in the eccentricity would eventually convert all orbits into hyperbolas, or even worse would decrease e below zero. Finally, a secular term in the inclination would lead to an end over end tumble of the orbital plane, an effect not expected in a system which is symmetric about the z axis. The secular effects we did find are all real, and do not violate the integrity of the final result.

In fact, there are two very important applications of these J_2 secular terms. Low orbit weather satellites (the Nimbus series) and earth resources satellites (the Landsats) are launched into orbits where the node regresses at precisely 360° /year, or slightly less than one degree per day. As shown in Figure 5.3, this locks the satellite's orbital plane into a constant relationship to the sun, and such satellites are called sun – synchronous. Since such satellites are usually launched around midday local time, this ensures that forever afterwards half the satellite's orbit will be in sunlight, and that the satellite will not end up orbiting over the earth's 'twilight zone'. It also has the effect of ensuring that all photographs taken by the satellite are taken at the same local time for the area being photographed. This is a great aid in photointerpretation, since the user does not have to deal with the effects of different solar illumination. In fact, the different satellites of the Landsat series are known by their time of day: the 10 AM satellite, the 2 PM satellite, and so forth. Equating the nodal secular rate to the desired value supplies one relation between a, e, and i. If we

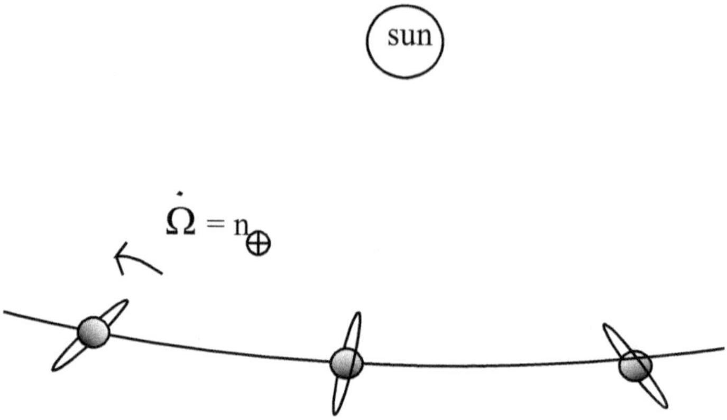

Figure 5.3: The Sun – Synchronous Orbit

wish our sun – synchronous satellite orbit to be circular, then we find a relation between a and i. It is extremely convenient that for the earth, at least, sun – synchronous orbits are nearly polar. This makes possible an orbit combining global coverage and constant illumination angle in one satellite.

The second application concerns the rate of advance or regression of the line of apsides. Setting $d\omega/dt$ to zero, we find that the orbit does not rotate when

$$\frac{5}{2}\sin^2 i - 2 = 0 \tag{5.65}$$

This yields the solution

$$i_c = \sin^{-1}\left(\frac{2}{\sqrt{5}}\right) \tag{5.66}$$

The two angles $i_c = 63.4349^o$ or $i_c = 116.5650^o$ are called the *critical inclinations.* They do not depend on the value of J_2 for the planet being orbited. An orbit at these inclinations will not rotate in its own plane. This type of orbit is used extensively by the Russians for its high eccentricity Molyniya communications satellites. Since the apogee must be in the northern hemisphere to be useful for high latitude communications use, the Russians need $\omega = 270^o$. Furthermore, if ω changes with time, apogee would migrate into the southern hemisphere, and the satellite would be useless for an extended period of time. By launching these satellites at the precise value of the critical inclination, this problem can be avoided. We will return to this problem in a later chapter, because there are further interesting dynamics operating in this situation.

Finally, notice that there is now some confusion as to what the period of the orbit is. We have already introduced the period from perigee to perigee, the anomalistic period. However, the perigee is moving with time, and the *siderial period* of the satellite (its period with respect to the stars) is different. Since the perigee advances for low inclinations, the anomalistic period will be slightly greater than the siderial period. The nodal period is the time interval which elapses between equator crossings, and is used to construct the satellite's ground track. Since the node regresses, the nodal period of the satellite is the shortest of the three periods.

5.7 Motion about an Oblate Planet

We are now just about prepared to derive the perturbation solution for satellite motion about an oblate planet. The expansion of R_2 can be factored to yield the somewhat simpler form

$$\begin{aligned} R_2 &= \frac{\mu J_2 R_\oplus^2}{a^3}\left[\left(-\frac{1}{4}+\frac{3}{4}\cos^2 i\right)\left(1+\frac{3}{2}e^2\right.\right. \\ &+ \left. 3e\cos M+\frac{9}{2}e^2\cos 2M\right)+\left(\frac{3}{4}-\frac{3}{4}\cos^2 i\right) \\ &\times \left(-\frac{3}{2}e\cos(M+2\omega)+(1-\frac{5}{2}e^2)\cos(2M+2\omega)\right. \\ &+ \left.\left.\frac{7}{2}e\cos(3M+2\omega)+\frac{17}{2}e^2\cos(4M+2\omega)\right)\right]+\mathcal{O}(e^3) \end{aligned} \tag{5.67}$$

Alternately, we are free to proceed to do perturbation theory using the canonical Delaunay variables. Since the classical elements are given by

$$a=\frac{L^2}{\mu}, \quad e=\sqrt{1-G^2/L^2}, \quad \cos^2 i = H^2/G^2 \tag{5.68}$$

$$l=M, \quad g=\omega \quad h=\Omega \tag{5.69}$$

and since we need the perturbing potential, the negative of the disturbing function, the Hamiltonian for J_2 perturbations in the Delaunay elements is

$$\begin{aligned} K_2 &= -\frac{\mu^2}{2L^2} \\ &- \frac{\mu^4 R_\oplus^2 J_2}{L^6}\left[\left(-\frac{1}{4}+\frac{3}{4}\frac{H^2}{G^2}\right)\left(\frac{5}{2}-\frac{3}{2}\frac{G^2}{L^2}\right.\right. \\ &+ \left. 3\sqrt{1-G^2/L^2}\cos l+\frac{9}{2}\left(1-\frac{G^2}{L^2}\right)\cos 2l\right) \\ &+ \frac{3}{4}\left(1-\frac{H^2}{G^2}\right)\left(-\frac{3}{2}\sqrt{1-G^2/L^2}\cos(l+2g)\right. \end{aligned} \tag{5.70}$$

$$
\begin{aligned}
&+ \left(-\frac{3}{2}+\frac{5}{2}\frac{G^2}{L^2}\right)\cos(2l+2g)+\frac{7}{2}\sqrt{1-G^2/L^2}\cos(3l+2g) \\
&+ \frac{17}{2}\left(1-\frac{G^2}{L^2}\right)\cos(4l+2g)\Bigg]+\mathcal{O}(e^3)
\end{aligned}
$$

However, we will not derive the complete solution in this section. Rather, we will explore the form of the solution. The proper place for such a solution is in computer code to evaluate the result. Hand calculation is far too cumbersome.

Let us begin with the semimajor axis, and see if we can obtain some general information on the form of the complete solution. The Lagrange Planetary equation for a becomes

$$
\begin{aligned}
\frac{da}{dt} &= \frac{2\mu J_2 R_\oplus^2}{na^4}\left[\left(-\frac{1}{4}+\frac{3}{4}\cos^2 i\right)(-3e\sin M\right. \\
&- 9e^2\sin 2M)+\left(\frac{3}{4}-\frac{3}{4}\cos^2 i\right)\left(\frac{3}{2}e\sin(M+2\omega)\right. \\
&- (2-5e^2)\sin(2M+2\omega)-\frac{21}{2}e\sin(3M+2\omega) \\
&- \left.\left. 34e^2\sin(4M+2\omega)\right)\right] \qquad (5.71)
\end{aligned}
$$

Substituting the two body solution on the right side makes a, e, i, and ω constant, while $M = nt + M_o$. The resulting expression can then be integrated with respect to time to give

$$
\begin{aligned}
a(t) &= a_o - \frac{2\mu R_\oplus^2 J_2}{n^2a^4}\left[\left(-\frac{1}{4}+\frac{3}{4}\cos^2 i\right)\left(-3e\cos M\right.\right. \\
&- \left.\frac{9}{2}e^2\cos 2M\right)+\left(\frac{3}{4}-\frac{3}{4}\cos^2 i\right)\left(\frac{3}{2}e\cos(M+2\omega)\right. \\
&- (1-\frac{5}{2}e^2)\cos(2M+2\omega)-\frac{7}{2}e\cos(3M+2\omega) \\
&- \left.\left.\frac{17}{2}e^2\cos(4M+2\omega)\right)\right]\Bigg|_0^t \qquad (5.72)
\end{aligned}
$$

As we would expect after the last section, the perturbations in the semimajor axis are purely periodic. The elements appearing on the right side of (5.72) are to be considered to be the constant initial conditions, $a = a_o, e = e_o$, etc. These are often referred to as mean elements, particularly when they are chosen to be the mean values when the periodic terms are averaged out.

Another simple one is the inclination. The Lagrange Planetary equation for the inclination is

$$
\frac{di}{dt} = \frac{\cot i}{na^2\sqrt{1-e^2}}\frac{\partial R}{\partial\omega} \qquad (5.73)
$$

since the J_2 harmonic is symmetric about the z axis, and is therefore free of Ω. After taking the partial derivative, substituting the two body solution, and integrating with respect to time, the result is

$$\begin{aligned} i(t) &= i_o - \frac{\mu R_\oplus^2 J_2 \cot i}{2n^2 a^5 \sqrt{1-e^2}} \left[-\frac{3}{4} \sin^2 i \, (3e \cos(M + 2\omega) \right. \\ &+ \left(1 - \frac{5}{2}e^2\right) \cos(2M + 2\omega) + \frac{7}{3} e \cos(3M + 2\omega) \\ &+ \left. \left. \frac{17}{4} e^2 \cos(4M + 2\omega) \right) \right] \Bigg|_0^t \end{aligned} \tag{5.74}$$

This is also purely periodic in nature.

However, the expected nasty surprises await if we try to complete this theory using the classical elements. The Lagrange Planetary equations for e, ω, and M contain the term $1/e \; \partial R / \partial e$. This simple term reduces our beautiful, hard won second order expansion of R_2 to a result that is only of the zeroth order in the eccentricity! In fact, if we wanted a result correct to the second order in e, we would have to expand the disturbing function to the *fourth order.* Furthermore, the terms which were of order 1 in the eccentricity are of order -1 in the result: the eccentricity appears in the denominator of some terms. This is the much feared singularity for zero eccentricity, a problem which is eliminated by converting R to the variables h and k. Also, a similar problem appears in the node Ω and the inclination (above) for zero i. This difficulty is bypassed by changing variables in R to χ and ψ.

The elements M, ω, and Ω will have both secular and periodic terms in the resulting series solutions. Perturbation solutions are only useful within a computer program designed to actually calculate the solution.

5.8 Air Drag II, Eccentricity Decay

We have already seen that the force form of Lagrange's equations is well adapted to studying air drag. In this section we will extend our earlier results on the behavior of the semimajor axis to consider the behavior of the eccentricity. First, the Lagrange equation for the eccentricity

$$\frac{de}{dt} = \frac{\sqrt{1-e^2} \sin \nu}{na} a_r + \frac{\sqrt{1-e^2}}{na^2 e} \left(\frac{a^2(1-e^2)}{r} - r \right) a_s \tag{5.75}$$

is not quite as singular as it looks. Take the critical factor $a^2(1-e^2)/r - r$ from this equation, and substitute for the radius vector

$$r = \frac{a(1-e^2)}{1 + e \cos \nu} \approx a - ae^2 - ae \cos \nu + ae^2 \cos^2 \nu \tag{5.76}$$

which is correct to $\mathcal{O}(e^3)$. We find

$$\frac{a^2(1-e^2)}{r} - r \approx 2ae \cos \nu + ae^2 \sin^2 \nu \tag{5.77}$$

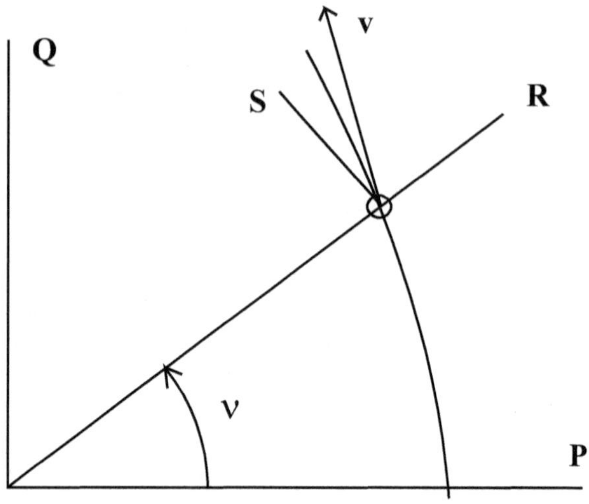

Figure 5.4: Velocity Coordinate Conversion

again neglecting terms of third order in e. The factor of e in the numerator eliminates the singularity in the Lagrange equation. (This makes it necessary to perform a second order expansion to obtain a first order result.) In fact, if we carry this level of approximation throughout, the factor $\sqrt{1-e^2} \approx 1$, and the Lagrange equation for the eccentricity becomes

$$\frac{de}{dt} = \frac{\sin\nu}{na} a_r + \frac{1}{na}\left(2\cos\nu + e\sin^2\nu\right) a_s + \mathcal{O}(e^2) \tag{5.78}$$

which is non–singular. (This approach *doesn't* work with the argument of perigee. The argument of perigee really does become undefined as $e \to 0$, so h and k must be used.) We will use this form to study the behavior of the eccentricity under air drag.

Now, the air drag acceleration is

$$\mathbf{a}_d = -\frac{1}{2} B^* \rho(r) v \mathbf{v} \tag{5.79}$$

and we still need convenient expressions for the scalar speed v, the velocity vector $\mathbf{v}$ expressed in the $\mathbf{R}$, $\mathbf{S}$ frame, and the air density $\rho(r)$. Begin with the velocity vector. The two body velocity vector can be written as

$$\mathbf{v} = \sqrt{\frac{\mu}{p}}\left(-\sin\nu\mathbf{P} + (e + \cos\nu)\mathbf{Q}\right) \tag{5.80}$$

where (see Figure 5.4) the $\mathbf{P}$ unit vector points towards perigee, and $\mathbf{Q}$ is along the latus rectum. Now, by inspection we have

$$\begin{aligned} \mathbf{P} &= \cos\nu\mathbf{R} - \sin\nu\mathbf{S} &\quad (5.81) \\ \mathbf{Q} &= \sin\nu\mathbf{R} + \cos\nu\mathbf{S} &\quad (5.82) \end{aligned}$$

Substituting then gives, after simplifying and replacing $p = a(1-e^2) \approx a$ to order e^2

$$\mathbf{v} \approx \sqrt{\frac{\mu}{a}}\left(e\sin\nu\mathbf{R} + (1+e\cos\nu)\mathbf{S}\right) \tag{5.83}$$

We then easily find

$$v = |\mathbf{v}| = \sqrt{\frac{\mu}{a}}\left\{1 + 2e\cos\nu + \mathcal{O}(e^2)\right\}^{1/2} \tag{5.84}$$

by taking the norm of $\mathbf{v}$. Using a binomial expansion for the square root, $\sqrt{1+\epsilon} \approx 1 - \epsilon/2$ reduces this to

$$v \approx \sqrt{\frac{\mu}{a}}\left(1 + e\cos\nu\right) \tag{5.85}$$

again neglecting terms of order e^2.

Finally, we need to expand the air density. Since the scale height h of the atmosphere is only 6–10 km, it is quite likely that the altitude difference between apogee and perigee will be comparable to this. The air density could then differ substantially between apogee and perigee. Using an exponential atmosphere we have

$$\rho(r) = \rho_o \exp\left\{-\frac{r - R_\oplus}{h}\right\} \tag{5.86}$$

where $R_\oplus$ is the radius of the earth. Using our approximation for the radius vector, this can be separated as

$$\begin{aligned} \rho(r) &\approx \rho_o \exp\{-(a - R_\oplus - ae\cos\nu)/h\} \\ &= \rho_o \exp\{-(a - R_\oplus)/h\}\exp\{ae\cos\nu/h\} &\quad (5.87) \\ &= \rho(a)\exp\{ae\cos\nu/h\} \end{aligned}$$

The quantity $\rho(a)$ should be nearly constant on a timescale of a few orbits, too short for the decay of the semimajor axis to be apparent. If we assume that the eccentricity is *quite* small, so that $ae << h$, then

$$\rho(r) \approx \rho(a)\left(1 + \frac{ae\cos\nu}{h}\right) \tag{5.88}$$

as usual to within terms of order e^2. If the eccentricity is larger, then we will have to keep more terms in the Taylor's series for the exponential function.

Inserting all these approximations into the expression (5.79) for the air drag gives, after again discarding higher order terms in the eccentricity

$$\begin{aligned} \mathbf{a}_d &= -\frac{1}{2}B^*\rho(a)\frac{\mu}{a}\{(e\sin\nu)\,\mathbf{R} \\ &+ \left(1+2e\cos\nu+\frac{ae\cos\nu}{h}\right)\mathbf{S}\} \end{aligned} \tag{5.89}$$

From this expression the components a_r and a_s are easily extracted. Substituting into the non–singular eccentricity Lagrange equation then gives

$$\begin{aligned} \frac{de}{dt} &= -\frac{1}{2}B^*\rho(a)\frac{\mu}{na^2}\,(2\cos\nu+2e\sin^2\nu+4e\cos^2\nu \\ &+ \frac{2ae}{h}\cos^2\nu\Big) \end{aligned} \tag{5.90}$$

Now, there are two ways we can tackle the above differential equation. If we proceed as usual, we would substitute the two–body behavior for $\nu(t)$ into (5.90), expand again, and then integrate. This approach is still valid, and will yield what are termed the *short period periodic perturbations* in the eccentricity. On the other hand, if the effects of air drag are very small, then the eccentricity will not change very much over one orbit. If this is true, the *method of averaging* can be used to find the long period behavior of the eccentricity. The basic idea is that the short period perturbations are periodic, and do not lead to any cumulative effect. If the eccentricity is nearly constant over one orbit, the periodic coefficients do not produce much of an effect, and we might as well consider only their average values in the differential equation. We then *average* the coefficients in the above equation, holding the eccentricity constant over one orbit, in order to study longer term behavior. As

$$\langle\cos\nu\rangle = \frac{1}{2\pi}\int_0^{2\pi}\cos\nu d\nu = 0 \tag{5.91}$$

and

$$\left\langle\cos^2\nu\right\rangle = \frac{1}{2\pi}\int_0^{2\pi}\cos^2\nu d\nu = \frac{1}{4\pi} = \left\langle\sin^2\nu\right\rangle \tag{5.92}$$

the averaged equation will not generate any periodic terms. (Properly, this should be a time average. But $\nu \approx M$ to first order.)

The averaged equation of motion for the eccentricity then is

$$\frac{d<e>}{dt} = -\frac{1}{4\pi}B^*\rho(a)\frac{\mu}{na^2}\left(3+\frac{a}{h}\right)<e> \tag{5.93}$$

This is of the form

$$\frac{d<e>}{dt} = -\alpha<e> \tag{5.94}$$

where α is a positive constant. Of course, this equation has the easily found solution

$$<e>=<e>_o\exp\{-\alpha(t-t_o)\} \tag{5.95}$$

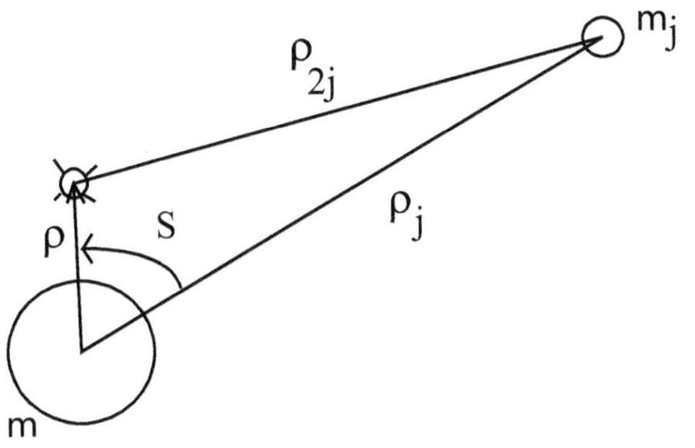

Figure 5.5: Third Body Perturbation Geometry

This says that the eccentricity decays towards zero, as advertised. Looking at the coefficient α, we notice that it is dominated by the term in a/h. For a low earth satellite this is at least $a/h \approx 1000$, and perhaps somewhat larger. By comparison, 3 is insignificant. So the decay of the eccentricity is almost entirely dominated by the change in air density between apogee and perigee.

As an alternative to the "method of averaging", equation (5.90) can be reduced to multiple angle form by eliminating the powers of the sine and cosine functions. The *secular* terms in this equation will simply give our averaged equation of motion. The other terms generate periodic perturbations. If we were to then follow our usual method and integrate the secular terms as $-\alpha t$, we would have the local slope of (5.95). However, this is a linear constant coefficient system (5.94) with a periodic forcing term, and can be solved correctly in closed form.

5.9 The Third Body Disturbing Function

For most of its 300 year history, the goal of celestial mechanics was to account for perturbations due to "third" masses. In section 4.6 we obtained the disturbing function for third body perturbations as

$$R = \mu_j \left(\frac{1}{\rho_{2j}} - \frac{\vec{\rho}_j \cdot \vec{\rho}}{\rho_j^3} \right) \tag{5.96}$$

where $\vec{\rho}_{2j}$ is the vector from the satellite to the perturbing mass, and $\vec{\rho}_j$ locates the third mass relative to the primary body. The first term in (5.96) is called the direct term, while the second term, called the indirect part, accounts for perturbations of the third mass on the primary body.

Introduce S, the angle between the vectors $\vec{\rho}$ and $\vec{\rho}_j$ as shown in Figure 5.5. Then, the indirect part can be written

$$R_{indirect} = -\mu_j \frac{\vec{\rho}_j \cdot \vec{\rho}}{\rho_j^3} = -\mu_j \left(\frac{\rho}{\rho_j^2} \right) \cos S \tag{5.97}$$

while the direct part becomes

$$R_{direct} = \frac{\mu_j}{\rho_{2j}} = \frac{\mu_j}{\rho_j} \left(1 - 2\alpha \cos S + \alpha^2\right)^{-1/2} \tag{5.98}$$

where $\alpha = \rho/\rho_j$ is the ratio of the two radii. The principal difficulty in third body perturbations is to expand the direct part of the disturbing function.

Historically, there have been two major approaches to this expansion. In the *Lunar Theory,* so named after the study of the motion of earth's moon, the quantity α is quite small. A direct expansion in powers of α is thus rapidly convergent. In the second method, used in *Planetary Theory,* the ratio α is not particularly small, and expansions in α are to be avoided at any cost. Both methods have been refined over literally hundreds of years of effort, and the expansion techniques have been brought to a high level of sophistication. We will *not* include all of that sophistication here.

In the "Lunar" case, where α is small, we can afford to expand the direct part (5.98) using the binomial theorem, to obtain

$$\begin{aligned}
(1 \quad - \quad & 2\alpha \cos S + \alpha^2)^{-1/2} \approx 1 + \frac{-1}{2} \left(-2\alpha \cos S + \alpha^2\right) \\
+ \quad & \frac{(-1/2)(-3/2)}{2!} \left(4\alpha^2 \cos^2 S - 4\alpha^3 \cos S + ...\right) \\
+ \quad & \frac{(-1/2)(-3/2)(-5/2)}{3!} \left(-8\alpha^3 \cos^3 S + ...\right) + \mathcal{O}(\alpha^4) \\
= \quad & 1 + \alpha \cos S + \alpha^2 \left(-\frac{1}{2} + \frac{3}{2} \cos^2 S \right) \\
+ \quad & \alpha^3 \left(-\frac{3}{2} \cos S + \frac{5}{2} \cos^3 S \right) + \mathcal{O}(\alpha^4) \\
= \quad & \sum_{k=0}^{\infty} \alpha^k P_k^0 (\cos S)
\end{aligned} \tag{5.99}$$

where the P_k^0 are the Legendre Polynomials. The second term in (5.99) above exactly cancels the indirect term (5.97), while the first term does not contain the coordinates of the satellite at all, so the "Lunar Theory" expansion of the disturbing function is given by

$$R = \frac{\mu_j}{\rho_j} \sum_{k=2}^{\infty} \left(\frac{\rho}{\rho_j} \right)^k P_k^0 (\cos S) \tag{5.100}$$

which begins with α^2 terms, and should converge very rapidly when α is small, as it would be for a close earth satellite perturbed by the moon or sun.

In the study of the motion of planets we cannot afford to expand in powers of α, which is typically about 0.3 to 0.7 within the major part of the solar system. Too many terms would need to be carried in order to ensure convergence. In the planetary case, a different form of expansion is used for the disturbing function. Begin by expanding the direct part of the disturbing function as a Fourier series in S

$$\left(1 - 2\alpha\cos S + \alpha^2\right)^{-1/2} = \frac{1}{2} b_{1/2}^{(0)}(\alpha) + \sum_{j=1}^{\infty} b_{1/2}^{(j)}(\alpha)\cos jS \tag{5.101}$$

The quantities $b_{1/2}^{(j)}$ are called Laplace coefficients, and are functions of α. The first two can be calculated in terms of the elliptic integral functions, while recursion relations exist to find the higher order coefficients.

To complete the expansion, the elements must be introduced. This introduces the elements M, ω, Ω, and i through S, and a, e, and ω through α. In the planetary case, this is done through expanding the Laplace coefficients using Newcomb operators. In Lunar theory few enough terms need be used that the brute force approach can be adopted for moons orbiting close to their planets. Earth's moon is an exception: it orbits at a relatively large distance. In either case, the end result is a series of the form

$$\begin{aligned} R &= \sum_{ijklmn} C(\alpha) e^{i'} e_p^{j'} \sin^{2k'} i \sin^{2l'} i_p \\ &\times \cos(iM + j\omega + k\Omega + lM_p + m\omega_p + n\Omega_p) \end{aligned} \tag{5.102}$$

where the six indices range from $-\infty$ to ∞, and the dependence on a (α) may not be explicit. This is called a Poisson series: the eccentricities and the sines of the inclinations enter as Taylor's expansion variables, while the mean anomalies, arguments of perigee, and nodes are part of a Fourier expansion. Of course, in the solar system eccentricities and inclinations are often small, and truncation of the series (5.102) based on this fact is greatly desired. We will not develop the full classical expansion techniques here. Rather, let us proceed to a simple example.

5.10 Third Body Perturbations

Let us consider the simplest example of the effects of third body perturbations. As shown in Figure 5.6, we assume that the orbit of the satellite is coplanar with the orbit of the moon, and we will take the moon's orbit to be circular. With respect to an inertial reference direction, the moon is at constant distance ρ_m from the earth, and is located by the angle $n_m t$. (Notice that this is the restricted three body problem.) The satellite is at distance ρ and angle $\nu + \omega$.

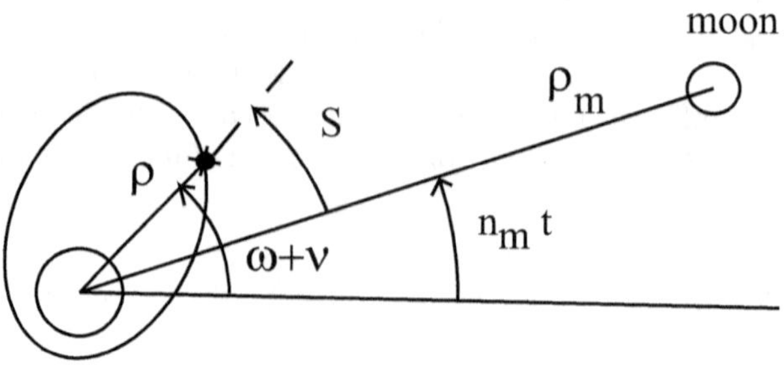

Figure 5.6: Coplanar Third Body Perturbations

The disturbing function for third body perturbations is, using the lunar theory form from the previous section, and truncating at the lowest order in ρ/ρ_m

$$R = \frac{\mu_m}{\rho_m^3}\rho^2\left(-\frac{1}{2}+\frac{3}{2}\cos^2 S\right) \tag{5.103}$$

For a close earth satellite, we would have $\alpha = \rho/\rho_m \approx 1/60$, while at geosynchronous orbit we still have $\alpha \approx 1/10$. Recall that (5.103) is correct to within terms of order α^3. To get this into a usable form, we will be forced to perform two expansions.

Our goal will be to obtain the lowest order results, without obscuring the physics of the problem in immense amounts of series expansions. So, we will expand only to the first order in the satellite eccentricity. The satellite radius then becomes, using the results of section 5.4,

$$\begin{aligned} \rho^2 &\approx a^2(1-e\cos M)^2 \\ &\approx a^2(1-2e\cos M)+\mathcal{O}(e^2) \end{aligned} \tag{5.104}$$

Similarly, the angle S between the moon and satellite is

$$S = \nu+\omega-n_m t \approx M+\omega-n_m t+2e\sin M \tag{5.105}$$

Expanding to the first order in e, we find

$$\begin{aligned} \cos^2 S &= \frac{1}{2}+\frac{1}{2}\cos 2S \\ &\approx \frac{1}{2}+\frac{1}{2}\cos(2M+2\omega-2n_m t) \\ &- e\cos(M+2\omega-2n_m t)+e\cos(3M+2\omega-2n_m t) \end{aligned} \tag{5.106}$$

Finally, inserting these two results into (5.103), and retaining only terms through the first order, we have

$$\begin{aligned} R \quad &\approx \quad \frac{\mu_m a^2}{\rho_m^3}\left\{\frac{1}{4}+\frac{3}{4}\cos(2M+2\omega-2n_m t)\right. \\ &- \quad \frac{1}{2}e\cos M+\frac{3}{4}e\cos(3M+2\omega-2n_m t) \\ &- \quad \left.\frac{9}{4}e\cos(M+2\omega-2n_m t)\right\}+\mathcal{O}(e^2) \end{aligned} \tag{5.107}$$

Now, we are ready to do some perturbation theory. Let us start with the Lagrange Planetary equation for the semimajor axis:

$$\begin{aligned} \frac{da}{dt} \quad &= \quad \frac{2\mu_m a}{n\rho_m^3}\left\{-\frac{3}{2}\sin(2M+2\omega-2n_m t)+\frac{1}{2}e\sin M\right. \\ &- \quad \left.\frac{9}{4}e\sin(3M+2\omega-2n_m t)+\frac{9}{4}e\sin(M+2\omega-2n_m t)\right\} \end{aligned} \tag{5.108}$$

Now, substitute the two body approximation $M = nt + M_o$, and integrate the result with respect to time. The result for the first order perturbation of the semimajor axis is

$$\begin{aligned} a(t) \quad &= \quad a_o+\frac{2\mu_m a_o}{n\rho_m^3}\left\{\frac{3}{4}\frac{1}{n-n_m}\cos(2M+2\omega-2n_m t)\right. \\ &- \quad \frac{e}{2n}\cos M+\frac{9}{4}\frac{e}{3n-2n_m}\cos(3M+2\omega-2n_m t) \\ &- \quad \left.\left.\frac{9}{4}\frac{e}{n-2n_m}\cos(M+2\omega-2n_m t)\right\}\right|_{t_o}^{t} \end{aligned} \tag{5.109}$$

This expression yields small periodic perturbations for small semimajor axis values, compared to the distance to the moon, ρ_m. However, there are three cases when the perturbations are (allegedly) infinite. If the satellite is in an orbit with the same semimajor axis as the moon, the denominator of the first term is zero, and this term becomes infinite. This is called the 1:1 resonance, and it is termed a zero order resonance since this term is independent of the eccentricity. The last two terms become infinite when the mean motion of the satellite is exactly twice that of the moon or 2/3 that of the moon. This occurs when $a \approx 0.629\rho_m$ for the 2:1 resonance, and outside the orbit of the moon for the 2:3 resonance. These are termed first order resonances, since the first power of the eccentricity appears in the offending term.

There are additional resonances when the expansion of the disturbing function is carried to higher orders. Additional first order resonances are generated if more terms are retained in the expansion in the ratio of the orbital radii. These resonances involve ratios of the mean motions of the form N:N-1, satellite to moon. Also, higher order expansion in the eccentricity would reveal second order resonances involving mean motion ratios of the form 3:1, 4:2, 5:3, etc, third order resonances of the form 4:1, 5:2, and so forth. In fact, if *you pick any ratio of the mean motions, there will be resonances of some order nearby.* These occur for any *rational* ratio of the mean motions.

Of course, there is nothing infinite about the forces on the satellite in the vicinity of a resonance. Proximity to a resonance means that the same relative configurations of the moon and satellite repeat in the same order: at the 2:1 resonance the same relative satellite/moon configuration repeats every 2 orbits of the satellite, or one orbit of the moon. This will give the small lunar gravitational force a chance to produce relatively large effects. The fundamental assumption of perturbation theory, that the perturbations will be small, has been violated in the vicinity of a resonance. It is not violated infinitely, but it is violated. The perturbation expansion itself informs us, in this case, when it runs into difficulty. Nonlinear resonance is one topic which we will take up in a later chapter.

5.11 Spin – Orbit Coupling

Very often, the coupling between the orbital motion and the rotational motion for an orbiting body is so small as to be completely negligible. However, when an object becomes large enough and asymmetric enough, the rotational motion is influenced by the orbit. This is certainly true for a moon orbiting its parent planet, but satellites like the International Space Station are large enough that they cannot afford to ignore the "gravity gradient torque" produced by the close proximity of the earth. In extreme cases, the coupling can go the other way, as well, as two large, asymmetric objects orbit each other in very close proximity. Normally, however, the first coupling to surface is the influence of the orbit on the rotational motion of the satellite.

In this section consider the planar problem shown in Figure 5.7. The rigid body of mass M with moments of inertia A, B, C is spinning at rate $\dot{\psi}$ about the normal to the page, which we also assume is the orbital plane. The kinetic energy of rotation is then

$$T = \frac{1}{2}C\dot{\psi}^2 = \frac{1}{2C}p_\psi^2 \tag{5.110}$$

since the momentum is $p_\psi = \partial T/\partial\dot{\psi}$. Without any gravitational torques, equation (5.110) is the entire Hamiltonian, and we immediately learn that p_ψ and $\dot{\psi}$ are constant. However, what happens when such coupling is present?

If the body is in an elliptical orbit about the mass m, by inspection of Figure 5.7 the direction cosines of the $\mathbf{R}$ vector are

$$l = \cos(\nu - 180^o - \psi) = -\cos(\nu - \psi) \tag{5.111}$$
$$m = \sin(\nu - 180^o - \psi) = -\sin(\nu - \psi) \tag{5.112}$$

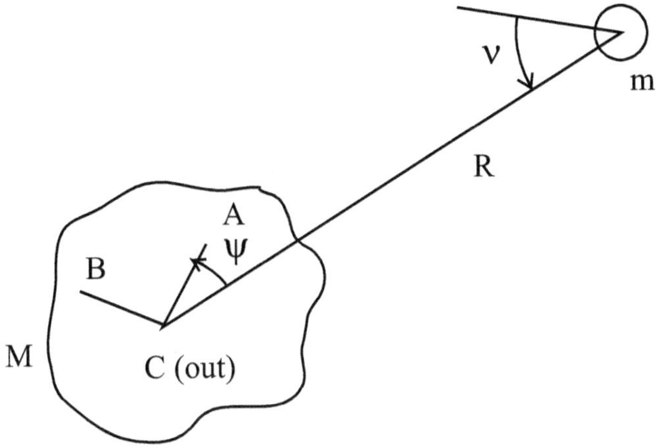

Figure 5.7: Spin – Orbit Coupling Geometry

Then, using the mass integral expansion of the potential energy from Chapter 4, problem 2, we have, (after dropping the Newtonian point mass term, since it will not contain the rotational coordinate ψ)

$$\begin{aligned} V &= -\frac{GMm}{4CR^3}\left\{\left(3\cos^2(\nu-\psi)-1\right)\left(1-\frac{A-B}{C}\right)\right. \\ &+ \left.\left(3\sin^2(\nu-\psi)-1\right)\left(1+\frac{A-B}{C}\right)\right\} \end{aligned} \tag{5.113}$$

where we have multiplied and divided by C, and dropped the third term of V since we are dealing with a planar problem. Now, define

$$\epsilon = \frac{A-B}{C} \tag{5.114}$$

For a nearly spherical mass ($A = B = C$), this will be a small parameter. The potential then becomes, after simplification

$$V = -\frac{GMm}{4CR^3}\left\{1 - 3\epsilon\cos 2(\nu-\psi)\right\} \tag{5.115}$$

We are now ready to insert the assumption that the rigid body is in a Keplerian orbit of small eccentricity. This means that

$$R^{-3} \approx a^{-3}(1-e\cos M)^{-3} \approx a^{-3}(1+3e\cos M) \tag{5.116}$$

using the binomial theorem, and

$$\begin{aligned} \cos 2(\nu - \psi) &\approx \cos(2\nu - 2\psi + 4e \sin M) \\ &\approx \cos 2(M - \psi) - 2e \cos(M - 2\psi) \\ &+ 2e \cos(3M - 2\psi) \end{aligned} \tag{5.117}$$

after using our normal expansion techniques, and discarding terms of order $\mathcal{O}(e^2)$. Inserting (5.116) and (5.117) into (5.115) and again expanding, we have

$$\begin{aligned} V &= \frac{GMm}{4Ca^3} \left\{ 3\epsilon \cos 2(M - \psi) - \frac{3}{2} e\epsilon \cos(M - 2\psi) \right. \\ &+ \left. \frac{21}{2} e\epsilon \cos(3M - 2\psi) \right\} \end{aligned} \tag{5.118}$$

A term involving only $\cos M$ has been dropped, since it has no bearing on the rotational problem. (It is the first orbit – to – spin coupling term.) Obviously, the lead term, which is free of the orbital eccentricity, will be the dominant perturbation term. However, it is instructive to carry the other two first order in e terms as well, even though they also contain ϵ and might be considered legitimate second order terms. They are also significant if e is not particularly small.

The rotational Hamiltonian then is

$$H = \frac{1}{2C} p_\psi^2 + V \tag{5.119}$$

where the entire potential energy is to be considered a perturbation on the kinetic energy term. The kinetic term alone leads to equations of motion

$$\dot{p}_\psi = -\frac{\partial H}{\partial \psi} = 0, \quad \dot{\psi} = \frac{\partial H}{\partial p_\psi} = \frac{p_\psi}{C} \tag{5.120}$$

These immediately integrate to give the zeroth order solution as

$$p_\psi = p_{\psi o}, \quad \psi = \psi_o + \dot{\psi} t \tag{5.121}$$

where the constant rotation rate is $\dot{\psi} = p_\psi / C$. Now, we are ready to do perturbation theory, using this solution as the initial guess.

The full equation of motion for the momentum is

$$\begin{aligned} \dot{p}_\psi &= -\frac{GMm}{4Ca^3} \{ 6\epsilon \sin 2(M - \psi) - 3e\epsilon \sin(M - 2\psi) \\ &+ 21 e\epsilon \sin(3M - 2\psi) \} \end{aligned} \tag{5.122}$$

Now, inserting the zero order solution for the behavior of ψ, and the usual two body solution for M, we have a form that can be integrated. The first order solution is

$$\begin{aligned} p_\psi(t) &= p_{\psi o} - \frac{GMm}{4Ca^3}\left\{-\frac{3\epsilon}{n-\dot\psi}\cos 2(M-\psi)\right. \\ &+ \left.\frac{3e\epsilon}{n-2\dot\psi}\cos(M-2\psi) - \frac{21e\epsilon}{3n-2\dot\psi}\cos(3M-2\psi)\right\}\Bigg|_0^t \end{aligned} \tag{5.123}$$

With the solution for the momentum in hand, the second of Hamilton's equations, $\dot\psi = p_\psi/C$ becomes

$$\begin{aligned} \dot\psi &= \frac{p_{\psi o}}{C} - \frac{GMm}{4C^2a^3}\left\{-\frac{3\epsilon}{n-\dot\psi}\cos 2(M-\psi)\right. \\ &+ \left.\frac{3e\epsilon}{n-2\dot\psi}\cos(M-2\psi) - \frac{21e\epsilon}{3n-2\dot\psi}\cos(3M-2\psi)\right\}\Bigg|_0^t \end{aligned} \tag{5.124}$$

This has solution

$$\begin{aligned} \psi(t) &= \psi_o + \dot\psi' t - \frac{GMm}{4C^2a^3}\left\{-\frac{3\epsilon}{2(n-\dot\psi)^2}\sin 2(M-\psi)\right. \\ &+ \left.\frac{3e\epsilon}{(n-2\dot\psi)^2}\sin(M-2\psi) - \frac{21e\epsilon}{(3n-2\dot\psi)^2}\sin(3M-2\psi)\right\}\Bigg|_0^t \end{aligned} \tag{5.125}$$

Here, $\dot\psi'$ includes the secular terms from the constant part of the momentum solution (13).

The solution for the momentum p_ψ and coordinate ψ are valid so long as the perturbations remain small. This means that the rotation rate $\dot\psi$ of the rigid body should not be anywhere near i) the orbit frequency n, ii) half the orbit frequency $n/2$, or iii) three halves of the orbital frequency $3n/2$. This is quite distressing, since a "horizon stabilized" satellite is going to have $\dot\psi = n$. Also, Nature herself insists on this case: every moon in the solar system for which we have data rotates at the same frequency as its orbital rate[1], and again we have the disastrous small divisor when $\dot\psi = n$. As a final example, the planet Mercury ($e \approx 0.3$) has a rotational rate which is precisely three halves that of its orbital period. Again, we have encountered a series of nonlinear resonances, this time exactly at the cases where both Nature and Mankind would like to operate. The solution above is invalid in these cases, but nothing catastrophic happens in the real world. We will return to this problem in a later chapter.

5.12 Von Ziepel's Method

We have now seen several examples of classical perturbation theory, where the original system is separated into the "solvable" and "unsolvable" parts, and the entire system is transformed into the variables which solve the solvable part.

[1]except for Hyperion at Saturn, which has chaotic rotation

The perturbation equations of motion can then be formulated, and (usually with much approximation and pain) integrated to give the first order perturbations. The first order solution can then, in principle, be substituted back into the perturbation equations to obtain higher order approximations. Higher order approximations are actually *usually necessary*. For example, the *second order* J_2 effects on a satellite orbit are of order 10^{-6}, which is basically the order of magnitude of the entire rest of the geopotential! So, there is little point to calculating first order perturbations for the rest of the geopotential expansion unless *second order* J_2 contributions are also included.

A more systematic method for performing higher order perturbation calculations is desirable. The oldest of these is Von Ziepel's method, and is based on the Hamilton – Jacobi equation

$$K = H + \frac{\partial S}{\partial t} \tag{5.126}$$

Assume that the original Hamiltonian is given by

$$H = H_o(q_i, p_i) + \epsilon H_1(q_i, p_i, t) \tag{5.127}$$

where the solvable part is H_o and the perturbation part is H_1, ϵ being a small parameter. In earlier chapters we used Hamilton – Jacobi theory to eliminate the zero order part in preparation for setting up the perturbation equations of motion. Von Ziepel decided to let the Hamilton – Jacobi equation do the entire job.

Assume that Hamilton's principal function can be expanded in a series in the small parameter ϵ

$$S = S_o + \epsilon S_1 + \frac{1}{2}\epsilon^2 S_2 + \ldots \tag{5.128}$$

As usual, the old momenta p_i are given by

$$p_i = \frac{\partial S}{\partial q_i} = \frac{\partial S_o}{\partial q_i} + \epsilon \frac{\partial S_1}{\partial q_i} + \frac{1}{2}\epsilon^2 \frac{\partial S_2}{\partial q_i} + \ldots \tag{5.129}$$

Now, insert this expansion into the Hamilton – Jacobi equation to find

$$\begin{aligned} K = 0 &= H_o(q_i, \partial S/\partial q_i) + \epsilon H_1(q_i, \partial S/\partial q_i) + \frac{\partial S}{\partial t} \\ &= H_o\left(q_i, \frac{\partial S_o}{\partial q_i} + \epsilon \frac{\partial S_1}{\partial q_i} + \frac{1}{2}\epsilon^2 \frac{\partial S_2}{\partial q_i} + \ldots\right) \\ &+ \epsilon H_1\left(q_i, \frac{\partial S_o}{\partial q_i} + \epsilon \frac{\partial S_1}{\partial q_i} + \ldots\right) + \frac{\partial S}{\partial t} \end{aligned} \tag{5.130}$$

This is all that is needed to form a second order theory.

Now, supposedly ϵ is small. We can expand both Hamiltonian terms in (5.130) about $\epsilon = 0$ to find

$$\begin{aligned} H_o(q_i, \partial S/\partial q_i) &\approx H_o(q_i, \partial S_o/\partial q_i) \\ &+ \sum_i \frac{\partial H_o}{\partial p_i}\left(\epsilon\frac{\partial S_1}{\partial q_i} + \frac{1}{2}\epsilon^2\frac{\partial S_2}{\partial q_i} + ...\right) \\ &+ \frac{1}{2!}\sum_i\sum_j \frac{\partial^2 H_o}{\partial p_i \partial p_j}\epsilon^2 \frac{\partial S_1}{\partial q_i}\frac{\partial S_1}{\partial q_j} + \mathcal{O}(\epsilon^3) \end{aligned} \tag{5.131}$$

and

$$\epsilon H_1 \approx \epsilon H_1(q_i, \partial S_o/\partial q_i) + \epsilon^2 \sum_i \frac{\partial H_1}{\partial p_i}\frac{\partial S_1}{\partial q_i} + \mathcal{O}(\epsilon^3) \tag{5.132}$$

All of the partial derivatives of H_o and H_1 are evaluated for $\epsilon = 0$. Finally, we need the final quantity

$$\frac{\partial S}{\partial t} = \frac{\partial S_o}{\partial t} + \epsilon\frac{\partial S_1}{\partial t} + \frac{1}{2}\epsilon^2\frac{\partial S_2}{\partial t} + \mathcal{O}(\epsilon^3) \tag{5.133}$$

and we are ready to substitute (5.131), (5.132) and (5.133) into the Hamilton – Jacobi equation.

This, of course, generates a terrible, terrible mess. However, the mess must still be true if $\epsilon = 0$, so we can separate these terms into their own equation. Now, if ϵ is not zero, but still infinitesimally small, the second order terms are negligible. The Hamilton – Jacobi equation would then contain only the zero and first order terms, and subtracting the zero order equation would require that the first order terms have their own equation. This argument can be repeated for the second and higher order terms, leading to a separate partial differential equation for each order of ϵ. We have imposed the condition that the expansion of the Principal Function S be *uniform* in ϵ. We obtain

$$\epsilon^0 \quad : \quad H_o\left(q_i, \frac{\partial S_o}{\partial q_i}\right) + \frac{\partial S_o}{\partial t} = 0 \tag{5.134}$$

$$\epsilon^1 \quad : \quad \sum_i \frac{\partial H_o}{\partial p_i}\frac{\partial S_1}{\partial q_i} + H_1\left(q_i, \frac{\partial S_o}{\partial q_i}\right) + \frac{\partial S_1}{\partial t} = 0 \tag{5.135}$$

$$\begin{aligned} \epsilon^2 \quad : \quad & \frac{1}{2!}\sum_i \frac{\partial H_o}{\partial p_i}\frac{\partial S_2}{\partial q_i} + \frac{1}{2!}\sum_i\sum_j \frac{\partial^2 H_o}{\partial p_i \partial p_j}\frac{\partial S_1}{\partial q_i}\frac{\partial S_1}{\partial q_j} \\ & + \sum_i \frac{\partial H_1}{\partial p_i}\frac{\partial S_1}{\partial q_i} + \frac{1}{2!}\frac{\partial S_2}{\partial t} = 0 \end{aligned} \tag{5.136}$$

and so forth for higher orders of ϵ. The zeroth order equation just says to go out and solve the solvable part of the system. Since we have already done this for the two body problem, we can bypass this step. The first order equation is a linear partial differential equation for the first order part S_1 of the generating function. In fact, every term in this equation can be calculated from known quantities *except* the quantities $\partial S_1/\partial q_i$ and $\partial S_1/\partial t$. This portion of the generating function will generate the first order perturbation solution once the coordinate transformations it dictates are found. Next, equation (5.136) becomes completely known with the exception of the term $\partial S_2/\partial q_i$; again, it is a linear partial differential equation. Further simplifications usually occur at the second and higher orders, since H_1 is rarely a function of the momenta p_i.

We also have the freedom to pick a new Hamiltonian K as something other than $K = 0$. Von Ziepel's own favorite was to choose

$$K = -\frac{\mu^2}{2L'^2} \tag{5.137}$$

where L' was the first new momentum. This choice for K appears on the right side of equation (5.134), while the right sides of equations (5.135), (5.136), etc remain zero. What we are then doing is performing a canonical transformation which maps the real system *onto the two body problem.* This is usually termed the elimination of short period terms. Alternately, in the case of a resonance problem, the new Hamiltonian K can be chosen to include both the two body part and the resonant terms, in which case the Von Ziepel transform "solves" the non – resonant part of the problem, leaving the resonance solution to be done as a final step.

Actually, there is no way to make a high order perturbation solution humane. Using modern computer algebra packages, however, can make incredible efforts possible. A standard test for such packages is to reproduce Delaunay's perturbation solution for the motion of the moon. Containing over 500 terms, it served as the basis for the predictions of the moon's motion in the American Ephemeris until recently. It took Delaunay *over 20 years* to perform this feat, and in the process, in spite of multiple checks, he made about a dozen errors. Modern computer series packages can reproduce the entire effort and find those errors in only a few *hours.* Von Ziepel's method has been largely supplanted by a much more sophisticated method based on Lie series, by the work of Hori and Deprit.

5.13 The Method of Averaging

We have seen through our study of general perturbations that most perturbing forces lead to a host of small amplitude periodic terms in the final result. These terms, while necessary for an accurate prediction of the orbit, really contribute nothing to the question of long term stability and evolution of a dynamical system. However, since they are present, they do have the effect of greatly slowing the progress of a numerical integration. If our aim is to follow the long – term fate of a system, we are going to need to use every trick at our disposal to achieve anything meaningful in a finite amount of computer time. A good first step is to switch from simple physical coordinates to two body element equations of motion, as the Delaunay elements, the Lagrange planetary equations, or even Encke's method. Let the state vector of elements be $\mathbf{x}$. Then we are contemplating the numerical integration of

$$\dot{\mathbf{x}} = \mathbf{f}(\mathbf{x}, t) \tag{5.138}$$

This is a great improvement, since we are not using half of our significant figures merely integrating the two body part of the equations of motion, and this postpones the day when the slow (square root) growth of truncation and roundoff errors swallow any remaining significance in the output from integrating (5.138).

However, equations (5.138) still contain swarms of small amplitude periodic forcing terms. While the switch to elements enables us to take somewhat larger timesteps, since the elements don't change as drastically with time as the physical coordinates, we have not really made a dramatic improvement. Any numerical integrator will need a timestep short enough to follow the details of the changes in the forcing function $\mathbf{f}$, and the perturbing forces still change rapidly over the course of one orbit. We will still need to take hundreds of steps per orbit to follow the evolution of the system (5.138) accurately.

Most of this is simply the periodic terms, and we have little interest in wading through thousands of cycles of these looking for long period effects. So the question of eliminating the periodic stuff arises. Periodic terms will have the mean anomaly M of the satellite present in the argument of a sine or cosine. Often, if another orbiting object is the perturber, the mean anomaly M_p of the planet will also appear in various linear combinations with M. These periodic terms can be wiped from the perturbing force by *averaging* over one period of the orbit. Since

$$\int_0^{2\pi} \cos jM dM = \int_0^{2\pi} \sin jM dM = 0 \tag{5.139}$$

averaging over M will eliminate the "short period terms", hopefully leaving only the "long period terms" in the equations of motion. This should also be done for any planets, so in the method of averaging we integrate

$$< \dot{\mathbf{x}} >= \frac{1}{(2\pi)^2} \int_0^{2\pi} \int_0^{2\pi} \mathbf{f}(< \mathbf{x} >, t)\; dM\; dM_p \tag{5.140}$$

where $< \mathbf{x} >$ indicate averaged state variables. The integration can be done numerically if necessary, and then with the short period terms eliminated, our timestep can span many orbits at one jump.

It is possible to derive this method by more formal means. Consider the system Hamiltonian

$$H = -\frac{\mu^2}{2L^2} + \epsilon H_1(L, G, H, l, g, h, t) \tag{5.141}$$

and construct a canonical transformation to a new Hamiltonian

$$K = -\frac{\mu^2}{2L'^2} + \epsilon K_1(L', G', H', -, g', h', -) \tag{5.142}$$

The dashes – in K_1 indicate that all dependence on the mean anomaly l and the time t has been eliminated in the new system (5.142). Otherwise, H_1 and K_1 are identical. Usually the Von Ziepel method is used to construct this transformation. The resulting system (5.142) is free of short period terms, and can be numerically integrated (at worst) with very large timesteps. If desired, the short period terms can be restored by the inverse transformation to obtain the actual orbit.

However, the independent integration over the two mean anomalies in (5.140) hides an important assumption. If the two mean anomalies are not linked in any way, then we are assuming *that the system is not in resonance.* If the system was in a resonant state, then some linear combination of the system angles

$$j_1 M + j_2 M_p + j_3 \omega + j_4 \omega_p + j_5 \Omega + j_6 \Omega_p \tag{5.143}$$

would not rotate with time, but would oscillate about some fixed value. This is termed libration, and invalidates the separate averaging over the two mean anomalies. The averaging would need to be modified to include the linear dependence of the system angle variables, and then would be valid only within resonance.

5.14 References

Any orbital mechanics text from the eighteenth or early nineteenth centuries will contain a major section on expansions in elliptic motion. Often this constitutes an entire chapter, since these expansions of the solution to the two body problem are central to the ability to perform general perturbation studies. The methods used are so varied that no one particular approach was adopted here.

Much of the historical literature in orbital mechanics treats the problem of perturbations by "third" bodies. As such, there is an enormous literature on the expansion of the third body disturbing function. A complete discussion is given in Brouwer and Clemence, *Methods of Celestial Mechanics,* Academic Press, 1961, chapter 15. Simon Newcomb was the head of the Naval Observatory in Washington, D.C., and the full expansion of this disturbing function constitutes the first volume of the Proceedings of the Naval Observatory.

5.15 Problems

Problem 1. Using the approach of Problem 2, Chapter 2, construct a complete first order perturbation theory for the harmonic oscillator with a soft spring

$$H = \frac{1}{2m}p^2 + \frac{k}{2}q^2 + \epsilon q^3$$

where $\epsilon << k$.

Problem 2. In the limit of very large rotational speeds, the potential term in the Hamiltonian for the simple pendulum is negligible

$$H = \frac{1}{2l^2}p^2 - gl\cos\theta \approx \frac{1}{2l^2}p^2$$

Use the Hamilton – Jacobi technique to solve this approximate Hamiltonian, and then employ that transformation to construct a first order perturbation solution for the pendulum in the rotational region.

Problem 3. Rewrite Kepler's equation as

$$E - M = e\sin E$$

so $E - M$ is an odd periodic function of M, and can be expanded as a Fourier series. Show that the series takes the form

$$E = M + \sum_{k=1}^{\infty} \frac{2}{k}\mathcal{J}_k(ke)\sin kM$$

where

$$\mathcal{J}_k(ke) = \frac{1}{\pi}\int_0^{\pi} \cos(kE - ke\sin E)dE$$

is a Bessel function of the first kind. (Hint: integrate by parts!)

Problem 4. Obtain the J_4 contributions to the secular rates of ω, Ω, and M to the first order in e.

Problem 5. Expand the disturbing function for the C_{22}, S_{22} sectoral harmonic to the lowest order in the eccentricity (eg, to $\mathcal{O}(e)$), for a nearly polar orbit ($i = 90^o$). Do enough of the first order perturbation theory to show that you have resonant behavior near orbital periods which are rational multiples of the earth's rotation rate.

Problem 6. Consider the problem of Earth's moon perturbed by the sun, the latter object assumed to be in a circular orbit. The lowest order part of the disturbing function is

$$R = +\frac{\mu_\odot}{\rho_\odot^3}\rho^2\left\{\frac{1}{2}\left(3\cos^2 S - 1\right)\right\}$$

Using the cosine law of spherical trigonometry, the quantity $\cos S$ can be written as (see Figure 5.8)

$$\begin{aligned}
\cos S &= \cos(M_\odot - \Omega)\cos(\nu + \omega) + \sin(M_\odot - \Omega)\sin(\nu + \omega)\cos i \\
&= \frac{1}{2}(1 + \cos i)\cos(\nu - M_\odot + \omega + \Omega) \\
&+ \frac{1}{2}(1 - \cos i)\cos(\nu + M_\odot + \omega - \Omega)
\end{aligned}$$

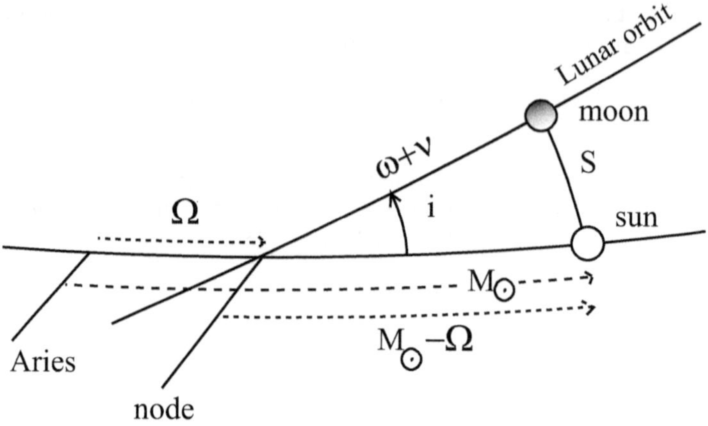

Figure 5.8: Solar perturbations on the Moon's orbit

Expand the disturbing function to the *zeroth* order in the eccentricity, and show that its secular part is

$$R_{sec} = \frac{\mu_\odot a^2}{a_\odot^3}\left\{\frac{3}{8}(1+\cos^2 i) - \frac{1}{2}\right\}$$

Calculate the secular rate of regression of the node of the Moon's orbit. The period of the moon is 27.8 days, while the period of the sun is 365.25 days. The inclination of the moon's orbit is small, about 4^0. Compare this period with the ancient Saros Cycle, used to predict eclipses.

Chapter 6

Nonlinear Resonance

6.1 Introduction

In performing our elementary perturbation theory for a satellite influenced by the gravity of the moon, we saw that nasty things happened when the mean motion of the satellite was the same as that of the moon, $n - n_m = 0$; or when the mean motion was twice that of the moon, $2n - n_m = 0$, or three halves that of the moon. Even if these relations are not obeyed exactly, if the satellite was close to either of these conditions the perturbations would be large. Since the fundamental assumption of perturbation theory is that the perturbations remain small, the result is suspect.

This condition is termed resonance, since the period of the satellite is a multiple of the period of the forcing term. Linear systems (like the forced harmonic oscillator) characteristically only have resonances when the period of the forcing term is the same as the period of the unperturbed system. This is termed the 1:1 resonance. However, nonlinear systems exhibit resonances at almost any rational multiple of the forcing frequency. Continuing the expansion of the lunar perturbations would have revealed additional resonances in the relations 3:2, 4:3, 5:4, all of order 1 in the eccentricity e, and of the form 3:1, 4:2, 5:3, of order e^2, and still further resonances of the form 4:1, 5:2, of order e^3, and so forth. There is a resonance within an infinitesimal distance from any initial orbit. However, most of these resonance structures are too narrow, and their timescales too long, to ever be noticed.

Nonlinear resonance also occurs for low earth satellites, driven by the geopotential. The bumps in the earth's gravity field rotate with the earth, once per siderial day. When the period of a satellite is close to a rational fraction of the siderial day, the satellite sees essentially the same bumps in the same order each day, and the perturbation has an opportunity to grow large. This leads to significant perturbations for satellites with periods near 14, 15, and 16 revolutions per day, typical low earth satellite periods. It also is the dominant effect in geosynchronous orbit, where the period of the satellite is one revolution per siderial day.

Finally, we will see that the problem of "small divisors" at resonance, periodic orbits, island structures on surfaces of section, and phase portraits of resonances are all different ways to look at the same thing.

6.2 The Simple Pendulum

Let us return to the simple pendulum one last time. This time we use the pendulum as a prototype of the behavior of a simple resonant system. The pendulum has a Hamiltonian function

$$H = \frac{1}{2l^2}p^2 - gl\cos\theta \tag{6.1}$$

where the canonical momentum $p = l^2\dot{\theta}$. This is a one degree of freedom problem with a conserved Hamiltonian. In fact, *any one degree of freedom problem with an integral is "reducible to quadratures"* in the old terminology. To see this, note that the Hamiltonian is the total energy E, and must be constant on any given trajectory. Solve (6.1) for p and replace it with its definition to obtain

$$\dot{\theta} = \pm\sqrt{2(E + gl\cos\theta)/l^2} \tag{6.2}$$

The variables can be easily separated to give

$$\pm\int_{\theta_o}^{\theta} \frac{d\theta}{\sqrt{2(E + gl\cos\theta)/l^2}} = \int_{t_o}^{t} dt = t - t_o \tag{6.3}$$

This is "reduction to quadratures": the dependence of θ on time has been reduced to the performing of two definite integrals, one of which we have already done. The old phrase means that, at worst, someone might have to perform the integral numerically.

In fact, the other integral in (6.3) is also known. Let $\theta = 2\phi$, and use the identity

$$\cos\theta = \cos 2\phi = 1 - 2\sin^2\phi \tag{6.4}$$

to obtain

$$\pm\int_{\theta_o}^{\theta} \frac{d\theta}{\sqrt{2(E + gl\cos\theta)/l^2}} = \pm\frac{2}{\sqrt{a}}\int_{\phi_o}^{\phi} \frac{d\phi}{\sqrt{1 - \sin^2\alpha\sin^2\phi}} \tag{6.5}$$

where

$$a = \frac{2E}{l^2} + \frac{2g}{l} \quad , \quad \sin^2\alpha = \frac{4g}{la} \tag{6.6}$$

This is the standard form for the incomplete elliptic integral of the first kind, $F(\phi \setminus \alpha)$, so equation (6.3) becomes

$$t - t_o = \pm\frac{2}{\sqrt{a}}\left(F(\phi \setminus \alpha) - F(\phi_o \setminus \alpha)\right) \tag{6.7}$$

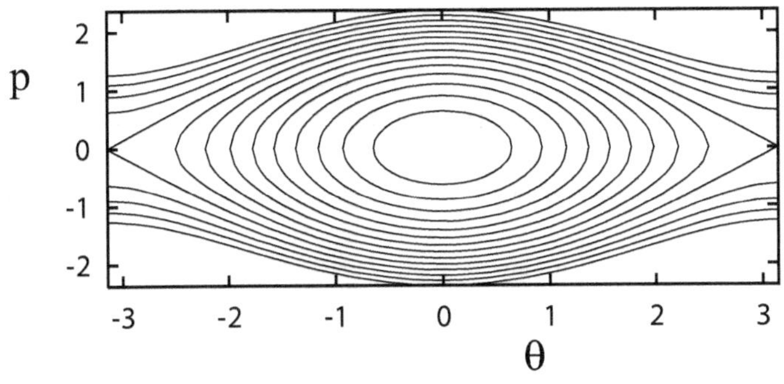

Figure 6.1: Phase Portrait for the Simple Pendulum

This is one form of the *closed form solution* to the motion of the simple pendulum.

Most resonance problems can be reduced to pendulum – like systems, and have solutions in terms of the various elliptic functions: elliptic integrals, Jacobi elliptic functions, or Wierstraussian elliptic functions. Abramowitz and Segun is, as usual, a good first place to go to become familiar with these functions if you need to use them. However, there is another technique available which can yield most of the information of the complete solution, and is much easier to implement. This is the method of phase portraits.

The Hamiltonian, equation (6.1), is a constant of the motion, so its' value does not change on any particular trajectory. It is a function of two variables alone: p and θ. A contour map of the function $H(p, \theta)$ is thus a map of *trajectories* on the p, θ plane, and is called a *phase portrait.* Figure 6.1 shows a phase portrait for the simple pendulum. The origin is surrounded by closed curves which represent trajectories of the mass in oscillatory motion. The origin, $p = 0, \theta = 0$, is obviously a stable equilibrium point. At $p = 0, \theta = \pi$ there is an unstable equilibrium, which correspond to trying to balance the pendulum vertically. The eigenvectors of this unstable saddle point extend across the figure and join to form separatrices, or eigencurves on the phase portrait. These two curves separate oscillatory motion from rotational motion.

There are two rotational regions, corresponding to the two directions in which the pendulum can rotate. In fact, since $p = l^2\dot{\theta}$, it is easy to add arrows to the phase portrait to note the direction in which the mass is moving. With the trajectories being level lines of the system Hamiltonian, and with the direction of motion being known from sign arguments on p, the only missing piece of information is how long it takes the system to move from one point to another on the phase portrait curve. This information cannot be obtained from the phase portrait, but must be found by other methods. The undulations in the rotational region of phase space become less and less pronounced as $|p|$ grows, since the effect of gravity becomes less noticeable for large rotational momentum. As $|p| \to \infty$, the contours become horizontal lines.

In the terminology of celestial mechanics, the oscillatory region is termed a *libration* region. This means that at least one variable is oscillating about a mean value. In a rotation region, all variables continually change with the same sign. The presence of both libration and rotational motion in the solution is a hallmark of resonance problems. Oscillatory functions like sine and cosine may suffice for the librational region, but cannot model the rotational motion. Polynomial functions of time (especially linear functions of time) may work in the rotational region, but cannot be used to model the librational motion. Elliptic functions, however, do both, and often show up in the closed form solution of resonant systems. Typically, perturbation theory fails in the vicinity of a resonance because *the zero order solution does not provide for the possibility of librational motion.* Since the zero order solution does not include it, a "small changes" perturbation solution is *not* going to be able to insert it. The only way to include it is to realize that the effect is large enough that it must be included in the zero order solution itself. Techniques for doing just this constitute the topic of this chapter.

6.3 Geosynchronous Orbit

In general, the zonal harmonics are symmetric about the earth's polar axis, so the rotational state of the earth is irrelevant. However, the sectoral and tesseral harmonics depend on the longitude of the satellite above the earth. As the earth rotates, this means that the disturbing function changes with time, with the period of the earth's rotation. If the satellite has a period close to a rational multiple of one siderial day, then we might suspect troubles with resonance terms. This is indeed the case.

Perhaps the most important case of geopotential resonance is the case of the geosynchronous equatorial satellite. In this case, the satellite is deliberately placed into an orbit as close to the one siderial day period as possible. Also, the satellite is put as close to zero eccentricity and inclination as is practical, so it will appear to be stationary seen from the earth. To see if this leads to any resonance problems, let us consider the perturbations on a geosynchronous orbit by the lowest order sectoral harmonic, the $n = 2, m = 2$ term. This is the largest term in the earth's gravity field which is not symmetric about the polar axis. The Legendre polynomial $P_2^2(\cos\theta) = 3\sin^2\theta$, so at the equator (where $\theta = 90^o$), we have $P_2^2 = 3$. Also, it makes little sense to include this term without also including the J_2 term, which is three orders of magnitude larger. The disturbing function, including $n = 2,\ m = 0$ and $n = 2,\ m = 2$ terms, is then

$$R = +\frac{\mu R_\oplus^2 J_2}{2r^3} + \frac{3\mu R_\oplus^2}{r^3}\left(C_{22}\cos 2\phi + S_{22}\sin 2\phi\right) \tag{6.8}$$

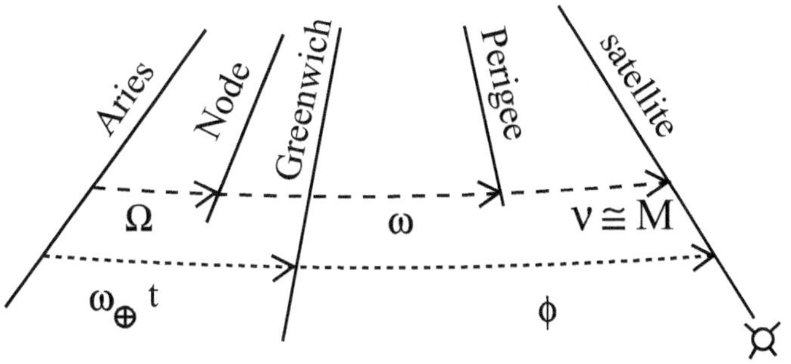

Figure 6.2: Orbital Longitude for a Geosynchronous Satellite

where ϕ is the geocentric longitude of the satellite.

Since we are only interested in the perturbations of satellites with very small eccentricities and inclinations, we will be content with the lowest possible order expansion of the disturbing function. If we assume that the orbit is perfectly circular and equatorial, then the true anomaly $\nu = M$, the mean anomaly, and the radius $r = a$, the semimajor axis. Furthermore, as shown in Figure 6.2,

$$\phi = \Omega + \omega + \nu - \omega_{\oplus} t \tag{6.9}$$

where $\omega_{\oplus}$ is the rotation rate of the earth, and ω and Ω are not really defined, since the eccentricity and inclination of the orbit are zero.

When this is substituted into the disturbing function, we obtain:

$$\begin{aligned} R &= \frac{\mu R_{\oplus}^2 J_2}{2r^3} + \frac{3\mu R_{\oplus}^2}{r^3}(C_{22}\cos 2(\Omega + \omega + M - \omega_{\oplus} t) \\ &+ S_{22}\sin 2(\Omega + \omega + M - \omega_{\oplus} t)) \end{aligned} \tag{6.10}$$

This can be simplified somewhat if we put

$$S_{22} = D_{22}\cos\delta_{22}, \quad C_{22} = D_{22}\sin\delta_{22} \tag{6.11}$$

This replaces the two constants C_{22} and S_{22} with two other constants D_{22} and δ_{22}, which we can interpret as an amplitude and a phase. Also, to $\mathcal{O}(e)$, we have $r = a$. The disturbing function now becomes

$$R = \frac{\mu R_{\oplus}^2 J_2}{2a^3} + \frac{3\mu R_{\oplus}^2}{a^3} D_{22}\sin(2(\Omega + \omega + M - \omega_{\oplus} t) + \delta_{22}) \tag{6.12}$$

Now, let us attempt a simple perturbation solution in classical elements. Since we have assumed that $e = 0$ and $i = 0$, only the semimajor axis a and mean anomaly M do not run into terrible troubles (all of which can be avoided by using the equinoctal variables). Looking at the Lagrange Planetary equation for a, we have

$$\frac{da}{dt} = +\frac{12R_\oplus^2 n}{a} D_{22} \cos\left(2(\Omega + \omega + M - \omega_\oplus t) + \delta_{22}\right) \tag{6.13}$$

remembering that $n^2 = \mu/a^3$. When we declare the two body elements on the right side to be approximately constant and integrate, the result is

$$a = a_o + \frac{6R_\oplus^2 n}{a(n - \omega_\oplus)} D_{22} \sin\left(2(\Omega + \omega + M - \omega_\oplus t) + \delta_{22})\right|_{t_o}^{t} \tag{6.14}$$

The mean anomaly perturbations come from

$$\frac{dM_o}{dt} = -\frac{2}{na}\frac{\partial R}{\partial a} \tag{6.15}$$

using the fact that $\partial R/\partial e = 0$ to sidestep the singularity in the Lagrange equation. This becomes

$$\frac{dM_o}{dt} = \frac{3\mu R_\oplus^2 J_2}{na^5} + \frac{18\mu R_\oplus^2 D_{22}}{na^5} \sin(2(\Omega + \omega + M - \omega_\oplus t) + \delta_{22} \tag{6.16}$$

Again assuming that the two body approximation can be used on the right side, we can integrate to obtain

$$\begin{aligned} M_o(t) &= M_o(t_o) + \frac{3\mu R_\oplus^2 J_2}{na^5}(t - t_o) \\ &- \left.\frac{9\mu R_\oplus^2 D_{22}}{na^5(n - \omega_\oplus)} \cos(2(\Omega + \omega + M - \omega_\oplus t) + \delta_{22})\right|_{t_o}^{t} \end{aligned} \tag{6.17}$$

The first term above is the J_2 secular correction to the mean motion, while the second term has an infinite coefficient at geosynchronous orbit.

These results are unacceptable. The perturbation in the semimajor axis is infinite when the orbit is precisely geosynchronous. However, this result cannot be correct. For one thing, we have assumed that the perturbation is *small*, and the results (6.14) and (6.17) cannot be trusted when the result is not small. To find out what is really happening, we are going to abandon the assumption that the perturbation is small, and try the problem again.

6.4 Geosynchronous Resonance

In the last section we saw that the perturbations in the semimajor axis due to the first zonal harmonic in the earth's gravitational field are not "small" near geosynchronous orbit. We cannot draw any other conclusions about the motion of the satellite near this orbit: perturbation theory has failed, and the result cannot be trusted. However, it is apparent that something "large" is happening near this orbit, and the resulting motion is not, in some sense, "close" to a two–body orbit. Near a resonance, we can use the perturbation theory results as a guide as to what is important. Terms which produce zero divisors are obviously going to need further attention. However, terms which only produce small effects in the perturbation expansion can be ignored while we tackle the main problem near the resonance. Alternately, they can be removed *before* we tackle the resonance problem, as we saw in the von Ziepel method.

So, since the orbit must not be sufficiently close to two body motion, we are going to return to the problems caused by the resonance term in the last section. We can write the "important" terms in the Hamiltonian for satellite motion near the 1:1 geosynchronous orbit as:

$$\begin{aligned} H &= -\frac{\mu^2}{2L^2} - \frac{\mu^4 R_\oplus^2 J_2}{2L^6} \\ &- \frac{3\mu^4 R_\oplus^2}{L^6} D_{22} \sin(2(l+g+h-\omega_\oplus t)+\delta_{22}) \end{aligned} \tag{6.18}$$

The first term is simply the two body Hamiltonian for orbital motion about the earth, in the Delaunay elements. The second term is the J_2 secular term, while the last term is the first zonal ($n = 2, m = 2$) harmonic disturbing function from the last section, transformed into these same elements, and added into the Hamiltonian after the necessary sign change. It is included in the Hamiltonian by virtue of the fact that it produces "infinite" results in perturbation theory, and therefore must be important near geosynchronous orbit. Everything else has been excluded for the same reason: other terms do not produce infinite results, and therefore must not be as important. We will study the system (6.18) as a new dynamical system in its own right, and see what we can learn about the solution.

The first thing we notice is that the system (6.18) has few apparent integrals of the motion. All three coordinates l, g, and h enter through the zonal term, so the conjugate momenta L, G, and H are not constant. However, the momenta G and H are missing, so the coordinates g and h are constants. (G and H show up when the expansion is carried to higher order in the eccentricity and inclination.) Also, time is present, so the Hamiltonian itself is not a constant of the motion.

However, all three coordinates and the time enter into the Hamiltonian only in the combination

$$\phi = l + g + h - \omega_\oplus t \tag{6.19}$$

If we were to chose ϕ as a new coordinate, and construct a canonical transformation, we might hope to eliminate two coordinates and time at one stroke, gaining three new integrals of the motion in the process. Notice, from Figure 6.2 of the last section that ϕ is simply the geocentric longitude of the satellite. Since ϕ includes all of the coordinate dependence of the Hamiltonian, the choice of the other two coordinates is not important, since they will not appear in the new system. So, chose new coordinates λ, ϕ, and k as:

$$\lambda = l, \quad \phi = l + g + h - \omega_{\oplus} t, \quad k = h \tag{6.20}$$

We have simply elected to keep the Delaunay coordinates $\lambda = l$ and $k = h$ as new coordinates in the new Hamiltonian.

Now, let the new momenta be Λ, Φ, and K, conjugate to λ, ϕ, and k. Using an F_2 generating function, the desired transform is given by

$$F_2 = \Lambda l + \Phi(l + g + h - \omega_{\oplus} t) + Kh \tag{6.21}$$

The other half of the transform relations yield the old momenta as

$$\begin{aligned} L &= \frac{\partial F_2}{\partial l} = \Lambda + \Phi \\ G &= \frac{\partial F_2}{\partial g} = \Phi \\ H &= \frac{\partial F_2}{\partial h} = \Phi + K \end{aligned} \tag{6.22}$$

The new Hamiltonian $\mathcal{K}$ is given by (remembering that $\partial F_2 / \partial t$ is not zero)

$$\begin{aligned} \mathcal{K} &= -\frac{\mu^2}{2(\Lambda + \Phi)^2} - \omega_{\oplus} \Phi \\ &- \frac{\mu^4 R_{\oplus}^2 J_2}{2(\Lambda + \Phi)^6} - \frac{3\mu^4 R_{\oplus}^2 D_{22}}{(\Lambda + \Phi)^6} \sin(2\phi + \delta_{22}) \end{aligned} \tag{6.23}$$

As expected, the number of integrals possessed by the system has dramatically risen.

In fact, λ, k, K, and time are missing from the new Hamiltonian $\mathcal{K}$, so Λ, K, k and the Hamiltonian $\mathcal{K}$ are all constants of the motion. The other variables Φ, ϕ the longitude of the satellite, and λ are not constants of the motion. However, our interest centers on ϕ especially, since this is a geosynchronous satellite. Also, while we will obtain a non-trivial equation of motion for the mean anomaly λ,

$$\dot{\lambda} = \frac{\partial \mathcal{K}}{\partial \Lambda} \tag{6.24}$$

This equation of motion does not contain λ on its right side. The right side contains only Φ, ϕ, and constants. So, the λ equation is decoupled from the rest of the system. We have essentially reduced the problem to a one degree of freedom dynamics problem and a definite integral, since all of the remaining interesting dynamics reside in Φ and ϕ. The λ motion can be obtained by an integration after this problem is solved.

Now, of the four integrals of the motion, $\mathcal{K}, \Lambda, K$, and k, two do not even appear in the Hamiltonian. The momentum Λ appears in the Hamiltonian, and is constant along an orbit. So, the one remaining constant of the motion is the

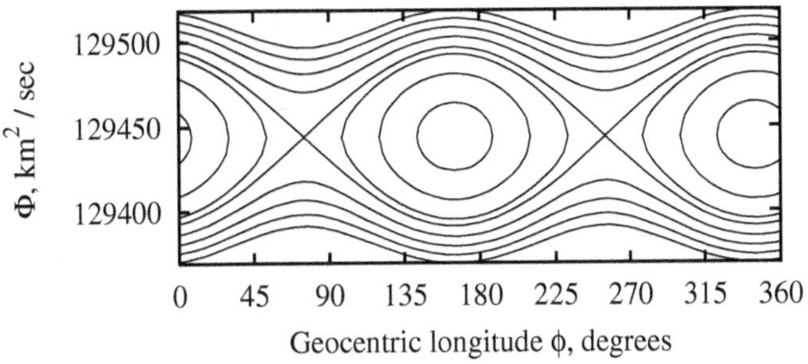

Figure 6.3: Phase Portrait for Geosynchronous Resonance

Hamiltonian itself, $\mathcal{K}$, and this contains both of the interesting coordinates Φ and ϕ. Since Φ and ϕ span a two dimensional space, and since we have one useful integral involving them, curves of constant Hamiltonian $\mathcal{K}$ in Φ, ϕ space are trajectories for the system. This is called the phase – portrait of the system. We will return to this technique shortly.

First, however, let us calculate the equations of motion. They are

$$\dot{\Phi} = -\frac{\partial \mathcal{K}}{\partial \phi} = +\frac{6\mu^4 R_\oplus^2 D_{22}}{(\Lambda + \Phi)^6} \cos(2\phi + \delta_{22}) \tag{6.25}$$

$$\begin{aligned} \dot{\phi} &= \frac{\partial \mathcal{K}}{\partial \Phi} = \frac{\mu^2}{(\Lambda + \Phi)^3} - \omega_\oplus + \frac{3\mu^4 R_\oplus^2 J_2}{(\Lambda + \Phi)^7} \\ &+ \frac{18\mu^4 R_\oplus^2 D_{22}}{(\Lambda + \Phi)^7} \sin(2\phi + \delta_{22}) \end{aligned} \tag{6.26}$$

The first two terms in equation (6.26) are dominant, and they can be recognized as $n - \omega_\oplus$. Thus, as expected, the rate of change of the longitude ϕ will be small near geosynchronous orbit. There will be small changes in the exact value of Φ which produces zero $\dot{\phi}$, depending on the value of Λ for a particular orbit. Equation (6.25) is dominated by the perturbation term, however, and it states that there are four values of ϕ which make the rate of change of Φ zero. This problem possesses four equilibrium points.

For the earth, $C_{22} = 1.57 \times 10^{-6}$ and $S_{22} = -0.897 \times 10^{-6}$, so the phase constant $\delta_{22} = -60.259^o$. Since $\cos(2\phi + \delta_{22}) = 0$ when the argument is 90° or 270°, the equilibrium points are located at longitudes $75.13^o, 165.12^o, 255.12^o$, and 345.12^o. These are located at longitudes near, respectively: India, the west coast of Australia, Colorado, and off the east tip of Africa.

Now, return to the idea of the phase portrait. The two important variables are Φ and ϕ, and we have one integral of the motion relating them, $\mathcal{K}$. Imagine plotting a trajectory on the Φ, ϕ plane. As the satellite's orbit changes, the Hamiltonian $\mathcal{K}$ must remain constant. Alternately, if we plot a contour map of the function $\mathcal{K}(\Phi, \phi)$, level lines of $\mathcal{K}$ *are actually trajectories across the* Φ, ϕ *plane.* This is not quite the complete solution to the problem, since we will have no information on how fast the satellite moves from one point to another. However, the major features of the motion can be read off from such a plot. All that is missing is the time information.

Figure 6.3 is a phase portrait for this system at a given, fixed value of Λ. Since $\Phi = G = L\sqrt{1-e^2}$, low eccentricity orbits (the only kind for which our equations of motion are valid) appear at the outer edge of the plot, where $\Phi \approx L$. Just inside the outer edge, where $e = 0$, appear our four equilibrium points. Two are stable, as can be seen from the fact that they are surrounded by closed contours of $\mathcal{K}$. The other two are unstable, and appear as saddle points. There are no other places around the equator where you can place a geosynchronous satellite and have it stay where it is put. At any other place, the satellite will begin to slowly drift away from its initial point. It will either librate about one of the stable equilibrium points, with an amplitude that could approach 180°, or it will circulate within either the inner or outer rotation regions. Stable equilibrium points surrounded by stable libration regions is the hallmark of nonlinear resonance. This structure changes with changes in the (constant) value of Λ.

It is this tendency to oscillate about a stable point, with a large amplitude and (usually) a long period which causes the failure of standard perturbation theory. This behavior is not a small change from two body motion. The two body assumption, from which we begin perturbation theory, contains only eastward drift if $a < a_{geo}$, or westward drift if $a > a_{geo}$. Two body motion does not have the large libration region centered on a_{geo}, and classical perturbation theory cannot reproduce it. The resonant term must be combined with the two body Hamiltonian to produce a new "zero order" problem for the analysis to be successful.

The large perturbations (with long period) produced by the first sectoral harmonic in geosynchronous orbit have a very practical effect on satellites. Since we do not want the large librational motion to show up, a stationkeeping system must be included in the satellite. In effect, we intend to park the vehicle on the side of a hill without setting the brake. The stationkeeping system is used to kick the satellite slightly 'uphill', after which it is allowed to drift 'up', slow down, stop, and begin to drift 'downhill' again. When the satellite reaches the boundary of its orbital window (set by antenna resolution on the earth's surface), it must be kicked uphill again. This results in a continual drain on stationkeeping fuel, and is a limiting constraint on the lifetime of a geosynchronous satellite. The satellite must characteristically expend about $\Delta v \approx 10$ m / sec / yr to maintain a desired longitude ϕ.

6.5 The Critical Inclination

When first encountered, there was considerable controversy surrounding the nature of the critical inclination. We have already seen in section 5.6 that at these inclinations the rate of change of the argument of perigee, $d\omega/dt$, becomes zero. This discussion is in the chapter on nonlinear resonance because the critical inclination is such a resonance. It is the 1:1 resonance between the two body frequency of the argument of perigee, namely zero, and the perturbation which produces the same zero frequency at the critical inclination. In his classic paper "The motion of an artificial satellite in the vicinity of the critical inclination"[1], Gen-Ichiro Hori showed that the secular disturbing function, including J_4, was

$$\begin{aligned}\mathcal{R}_{sec} &= -\frac{\mu R_\oplus^2 J_2}{2a^3(1-e^2)^{3/2}}\left(\frac{3}{2}\sin^2 i - 1\right) \\ &- \frac{5\mu R_\oplus^4 J_4 e^2}{8a^5(1-e^2)^{7/2}}\left(1 - 8\cos^2 i + 7\cos^4 i\right)\cos 2\omega \qquad (6.27)\end{aligned}$$

The secular part of the J_4 disturbing function can be found by calculating the constant term in its mean anomaly M Fourier series, just as was done for the J_2 term. Since one application will be the Russian Molynia satellites, which typically have eccentricities around $e \approx 0.7$, avoiding a low order expansion in e is very desirable.

If we try classical perturbation theory with the above, we are faced with one of two choices. If we begin with the assumption that the argument of perigee is constant, $\omega = \omega_0$, then we find that the secular rate depends on $\cos 2\omega_0$, which is a contradiction. The argument of perigee cannot both be constant and be changing linearly in time. Alternately, we can begin with the secular approximation through J_2, and write $\omega \approx \omega_0 + \dot{\omega}t$. This will produce perturbation solutions with $\dot{\omega}$ as a divisor, and this goes to zero at the critical inclination, just as if the latter were a resonance.

The problem can be converted to canonical form by changing variables to the Delaunay canonical set L, G, H, l, g, h, remembering to include the two body part of the Hamiltonian and reverse the sign on the disturbing function. The two body term in Delaunay elements is $\mathcal{H}_2 = -\mu^2/(2L^2)$. The conversion requires replacing

$$a = L^2/\mu, \quad \sqrt{1-e^2} = G/L, \quad \cos i = H/G$$

and

$$l = M, \quad g = \omega, \quad h = \Omega$$

and reversing the sign on the disturbing function. This leads to the canonical system

$$\mathcal{H} = -\frac{\mu^2}{2L^2} + \frac{\mu^4 R_\oplus^2 J_2}{2L^3G^3}\left(\frac{1}{2} - \frac{3}{2}\frac{H^2}{G^2}\right)$$

[1] Astronomical Journal volume 65, page 291, 1960

$$- \frac{5\mu^6 R_\oplus^4 J_4}{8L^5 G^5}\left(1-\frac{L^2}{G^2}\right)\left\{1-8\frac{H^2}{G^2}+7\frac{H^4}{G^4}\right\}\cos 2g \qquad (6.28)$$

As imposing as this looks, it is already in standard resonance form. The coordinates $l = M$ and $h = \Omega$ are missing from the Hamiltonian, so both L and H are constant. And while L is present, and so there is a non-trivial equation of motion for the mean anomaly l, this is completely decoupled from the remainder of the problem. So, (6.28) reverts to a one degree of freedom problem in G, $g = \omega$.

Hamilton's equations give the required equations of motion. They become

$$\begin{aligned} \dot{G} &= -\frac{\partial \mathcal{H}}{\partial g} \\ &= -\frac{5\mu^6 R_\oplus^4 J_4}{4L^5 G^5}\left(1-\frac{L^2}{G^2}\right)\left\{1-8\frac{H^2}{G^2}+7\frac{H^4}{G^4}\right\}\sin 2g \qquad (6.29)\end{aligned}$$

This immediately tells us that $\dot{G} = 0$ if

$$\omega = g = 0, \ \frac{\pi}{2}, \ \pi, \ \frac{3\pi}{2} \qquad (6.30)$$

The equation of motion for the argument of perigee is a little more complicated

$$\begin{aligned} \dot{g} &= \frac{\partial \mathcal{H}}{\partial G} \\ &= \frac{3\mu^4 R_\oplus^2 J_2}{2L^3 G^4}\left(\frac{5}{2}\frac{H^2}{G^2}-\frac{1}{2}\right) \\ &+ \frac{5\mu^6 R_\oplus^4 J_4}{8L^5 G^{12}}\{5G^6 - 77H^4L^2 - 7G^4(8H^2+L^2) \\ &+ 9G^2(7H^4+8H^2L^2)\}\cos 2g \qquad (6.31)\end{aligned}$$

The first term above is the usual argument of perigee secular rate $\dot{\omega}$. Notice it goes to zero when

$$5\frac{H^2}{G^2} - 1 = 5\cos^2 i - 2 = 0 \qquad (6.32)$$

which is the usual value for the critical inclination. The second term is more complicated, but is generally small. At one of the four values of $g = \omega = N\pi/2$, the second term becomes small, and a constant. The variable G is now a constant, and L and H already were constants. A small change in the "critical" inclination will absorb this term, and force $\dot{g} \to 0$. This makes $\dot{G} \to 0$, so there are 4 legitimate equilibria here. A phase portrait could be constructed, and it would look very much like several we have already seen, with two stable and two unstable equilibria, alternating in $\omega = g$.

The important question is which equilibria are the stable ones? The Russian Molynias use the one with $\omega = 3\pi/2$, which puts perigee as far south as it will go, and apogee as far to the north as possible. Hori was able to show that this is one of the *stable* equilibria. We will leave this as an exercise for the student.

6.6 Third Body Resonance

In section 5.10 we expanded the disturbing function for third body perturbations under the simplest possible assumptions, and found we had big difficulties on our hands for satellite periods close to twice the period of the moon and the same period as the moon. Higher order expansions would have produced all rational multiples of the period of the moon as places where classical perturbation theory fails. We have a method which might be acceptable on irrational period ratios, but which certainly fails for any rational period ratio. This is somewhat disconcerting, to say the least.

Let us study the 2:1 resonance in the earth – moon system. The term which caused so much trouble before is

$$R_{21} = -\frac{9}{4}\frac{\mu_m a^2 e}{\rho_m^3}\cos(M + 2\omega - 2n_m t) \tag{6.33}$$

This leads to an infinite perturbation term at the 2:1 resonance, where the divisor $n - 2n_m$ was zero. As we did in the last section, let us change the sign of this term, converting it to a potential energy, replace the classical elements with the Delaunay variables, and add in the two body Hamiltonian to find:

$$\mathcal{H} = -\frac{\mu^2}{2L^2} + \frac{9}{4}\frac{\mu_m L^4\sqrt{1 - G^2/L^2}}{\mu^2\rho_m^3}\cos(l + 2g - 2n_m t) \tag{6.34}$$

Near the 2:1 resonance we have included the two body problem and the trouble-making term, on the grounds that they are both important. We have excluded everything else, on the grounds that perturbation theory produces 'small' results for these terms, and therefore they are not important.

The Hamiltonian $\mathcal{H}$ appears to have no constants of the motion. All four coordinates and momenta are present, as is the time, so nothing appears to be constant. However, both coordinates and time only appear in only one place, in the combination $l + 2g - 2n_m t$. If we construct a transformation to a new set of variables where this is one of our new coordinates, we can hope to pick up some integrals. Pick as new coordinates

$$\sigma = l + 2g - 2n_m t, \quad \lambda = l \tag{6.35}$$

This implies an F_2 generating function

$$F_2 = \Lambda l + \Sigma(l + 2g - 2n_m t) \tag{6.36}$$

The other half of the transform relations give the old momenta as

$$L = \frac{\partial F_2}{\partial l} = \Lambda + \Sigma \tag{6.37}$$

$$G = \frac{\partial F_2}{\partial g} = 2\Sigma \tag{6.38}$$

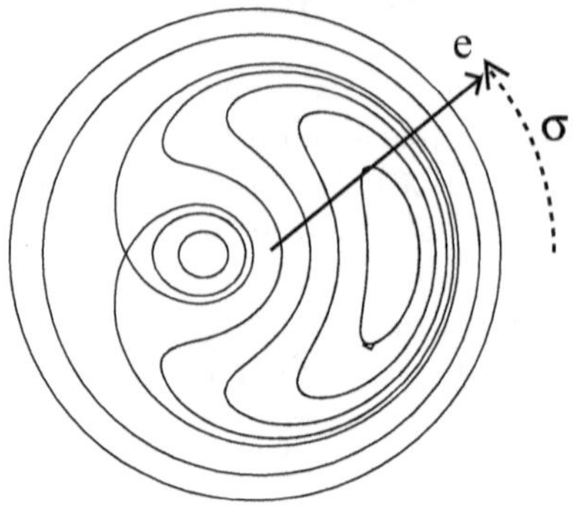

Figure 6.4: Level Curves of the Hamiltonian $\mathcal{K}$

And the new Hamiltonian is

$$\begin{aligned} \mathcal{K} &= -\frac{\mu^2}{2(\Lambda+\Sigma)^2} - 2n_m\Sigma \\ &+ \frac{9}{4}\frac{\mu_m(\Lambda+\Sigma)^4}{\mu^2\rho_m^3}\sqrt{1-\frac{4\Sigma^2}{(\Lambda+\Sigma)^2}}\cos\sigma \end{aligned} \tag{6.39}$$

The new Hamiltonian is constant, since time is missing, and in addition λ is missing, so the momentum Λ is also a constant of the motion. The equation of motion for λ is nontrivial, but is uncoupled from the rest of the problem. So, as with the geosynchronous case, a canonical transformation has reduced the problem to one degree of freedom system (Σ, σ) and a definite integral.

Also, just as with the geosynchronous case, level contours of the Hamiltonian $\mathcal{K}$ are trajectories on the e, σ plane. Figure 6.4 shows one such map, in the vicinity of the resonance. Outside the resonance, the trajectories on the e, σ plane are nearly circles, indicating that the eccentricity e (read Σ) is nearly constant, while the critical argument σ drifts at a nearly constant rate. This is exactly the behavior expected in the two body problem. Close to the resonance, however, the critical argument shows a region of libration. Again, this is a pendulum – like system, showing both rotation and libration. The stable periodic orbit has a gentleman's agreement with the moon, so that when it comes to apogee, it will be either 90° ahead or behind of the moon, while perigee points occur on the earth – moon line. The unstable periodic orbit (saddle point) has the opposite arrangement: every other apogee occurs near the moon, and the perturbations render this arrangement unstable. Since the two body problem can show only the rotation behavior, perturbation theory starting from the two body problem is doomed to failure.

6.7 Spin – Orbit Resonance

In section 5.11 we saw that the problem of planar rotation of a rigid body in an elliptical orbit led to resonance difficulties when the rotational rate of the satellite was the same as, half, and three halves of the orbital rate. As can be expected, there are additional resonances involving higher powers of the eccentricity e and the body mass factor $\epsilon = (A - B)/C$. However, these three resonances, and in particular the 1:1 resonance, sit exactly on cases of practical and natural interest. Virtually every moon in the solar system has a captured rotation rate: it keeps one face towards its parent planet. Horizon stabilized satellites, including the International Space Station, also operate at the 1:1 rotational resonance. Finally, the planet Mercury has wandered into the 3:2 rotational resonance with the sun.

This leads to three resonance problems: near the vicinity of the 1:1 resonance:

$$H_{1:1} = \frac{1}{2C}p_\psi^2 + \frac{3GMm\epsilon}{4Ca^3}\cos 2(M - \psi) \tag{6.40}$$

while near the 3:2 resonance:

$$H_{3:2} = \frac{1}{2C}p_\psi^2 + \frac{21GMme\epsilon}{8Ca^3}\cos(3M - 2\psi) \tag{6.41}$$

and near the 1:2 resonance:

$$H_{1:2} = \frac{1}{2C}p_\psi^2 - \frac{3GMme\epsilon}{8Ca^3}\cos(M - 2\psi) \tag{6.42}$$

Of these three, the first is of pressing practical and scientific interest, Mercury occupies the second case, and the third case is not observed in the solar system. (Apparently planets and moons are born with 'high' spin rates, and are spun down by tidal forces. Most objects have small orbital eccentricity, and are captured at the lowest order 1:1 resonance. Mercury has $e \approx 0.3$, and was captured at the 3:2 first order resonance. No natural object has made it past the 1:1 resonance to be captured into the 1:2 resonance.) In this section, we will examine the 1:1 resonance.

As is usual with the dropping of the non – resonant terms, the remaining system has all of its coordinate and time dependence isolated in one linear combination. Choose this as the new coordinate

$$Q = \psi - M \tag{6.43}$$

which immediately gives the generating function as

$$F_2 = P(\psi - M) \tag{6.44}$$

The old momentum is

$$p_\psi = \frac{\partial F_2}{\partial \psi} = P \tag{6.45}$$

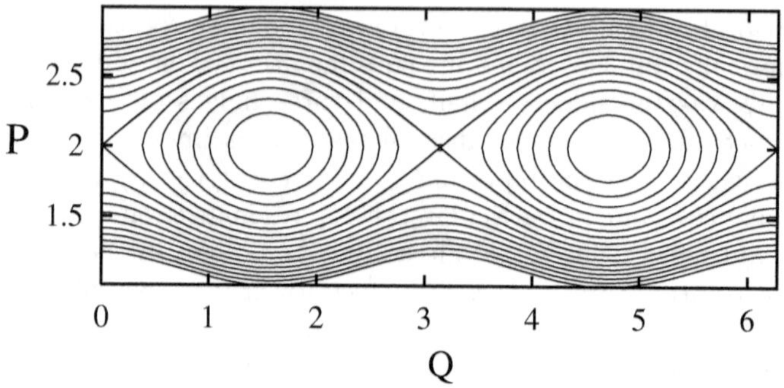

Figure 6.5: The 1:1 Spin – Orbit Resonance

and $\partial F_2/\partial t = -nP$, where n is the orbital mean motion. The new Hamiltonian then becomes

$$K_{1:1} = \frac{1}{2C}P^2 - nP + \frac{3GMm\epsilon}{4Ca^3}\cos 2Q \tag{6.46}$$

This is a one degree of freedom system, so there are no extra decoupled coordinates here. Equation (6.46) describes a pendulum – like system.

The equations of motion are

$$\dot{P} = -\frac{\partial K_{1:1}}{\partial Q} = \frac{3GMm\epsilon}{2Ca^3}\sin 2Q \tag{6.47}$$

$$\dot{Q} = \frac{\partial K_{1:1}}{\partial P} = \frac{1}{2C}P - n \tag{6.48}$$

The second of these states that $\dot{Q}$ is small if the satellite is near the resonant rotation rate (since $P = p_\psi$), while the first of these, in a sense, always predicts small changes. There are 4 equilibrium points, corresponding to P at the resonant value, and $\sin 2Q = 0$. The stable equilibrium always corresponds to pointing the *minimum moment of inertia axis* towards the parent planet, with the larger moment of inertia axis pointing along the velocity vector.

The Hamiltonian $K_{1:1}$ is also an integral of the motion, so phase – portrait analysis is possible. Such a diagram is shown in Figure 6.5, where it can be seen that the satellite is really indifferent as to which end of the minimum moment of inertia axis it points towards the planet. The resonant region is surrounded on either side by rotational regions, where the satellite rotates at a non – resonant rate, perturbed by the resonant term. This perturbation becomes smaller as we move further from the resonance. A gravity gradient satellite can execute large oscillations about the stable equilibrium, however, with very little damping. The Space Station will need active stability augmentation to help suppress these oscillations when they are excited by, for example, docking with a shuttle.

In the natural realm, the timescales to be considered are much longer. Furthermore, natural moons and planets are not really rigid bodies. In fact, they are much closer to being perfect fluids, with a thin scum of frozen stuff on their outside. The tidal interaction of the moon with its parent planet forces the moon to change its shape somewhat as it tries to rotate at a non – resonant rate (the planet raises a tide on the moon), and this supplies rotational damping. Since the stable equilibrium point represents the minimum energy value near the resonance, it can trap the decreasing initial rotation rate, freezing the object into synchronous rotation. This also makes the tide on the moon stationary, and the source of energy dissipation vanishes. Even after taking a major impact, the time required to damp back down to the equilibrium is usually 'short' (meaning only a few tens of million years). Earth's moon does not have any rotational "free libration" to within an accuracy of centimeters, an accuracy made possible by laser tracking targets left by the Apollo visits. (Notice that the argument in this paragraph can be made quite rigorous by applying the concept of Lyapunov stability).

6.8 Non – Linear Systems

By now you have hopefully noticed a certain similarity between various approaches to studying the structure of a non – linear system. Periodic orbits, found via boundary value problem techniques, locate certain special orbits and enable us to determine their stability properties by using Floquet theory. These same periodic orbits appear at the center of 'islands' on surface of section plots (the stable ones), or between the islands in a chain (the unstable periodic orbits). These same island chain structures then appeared in phase portraits of resonant systems, and again the centers and saddle points on the phase portraits correspond to stable and unstable periodic orbits.

They appear similar for a very simple reason: *they are virtually the same thing*. There are, however, some differences. The technique of ignoring the unimportant terms in the Hamiltonian leads to an integrable system in the vicinity of one resonance. This is the method we have used in this chapter. However, the terms we ignore typically generate resonances at some other point in the phase space. But resonances have finite width, so what happens when *two* resonances attempt to generate islands in the same region of space?

The problem of "two small divisors" has no known solution in general. If two terms lead to overlapping resonances, then there is no justification for discarding either term from the Hamiltonian, and that leaves us with too few integrals of the motion to use the phase portrait technique. Surface of section studies may show that isolated resonant island structures still exist, or they may reveal chaos. For chaos appears to begin when resonances start to overlap. This is termed Chirikov's conjecture. Since non – linear systems typically permit an infinite number of resonances, overlap is inevitable. The first order n : n-1 resonances in third body perturbations pack together as we approach the orbit of the third body (the 356:355 resonance is very close to that orbit), and three body systems typically degenerate into chaos near the orbit of the third mass. Geopotential resonances, on the other hand, are resonances with the earth's rotation, and their accumulation point is inside the earth itself. Chaos is not observed in this case.

Of course, it is the restrictive assumption of isolated resonances, as we have used it in this chapter, which breaks down. The surface of section method is always correct, although it can only be applied to a very restricted class of systems. Periodic orbit studies also are correct even when resonances overlap, although finding a good guess may be very difficult, if not impossible. None of these techniques can produce a full solution to a non – linear system, but they do permit us to glimpse the intricate beauty of the structure of phase space.

6.9 References

The basic method of dealing with an isolated resonance was discovered by Henri Poincaré. Boris V. Chirikov's work is found in Physics Reports, volume 52, No. 5, 1979.

6.10 Problems

Problem 1. Every time a navigation satellite is maneuvered, it becomes temporarily useless until its orbit is redetermined. So, it is important to avoid maneuvers, and nonlinear resonances can be used to do this.

Consider the influence of the $n = 4, m = 4$ geopotential term on the motion of a nearly polar $i \approx 90^0$ satellite orbiting with a period of nearly one half day. (Similar to a Global Positioning System (GPS) navigation satellite, except that their inclination is about 55^0.) Sketch, and show that the satellite's longitude is given by

$$\phi = \Omega - \omega_\oplus t$$

where $\omega_\oplus t$ is the siderial time at Greenwich, and that the colatitude is given by

$$\cos\theta = \sin(\omega + \nu) \approx \sin(\omega + M)$$

. Noting further that $P_4^4 = 105 \sin^4 \theta$, and setting

$$C_{44} \cos 4\phi + S_{44} \sin 4\phi = D_{44} \sin(4\phi + \delta_{44})$$

obtain the potential term in the form

$$\begin{aligned} V_{44} &= -\frac{105\mu R_\oplus^4 D_{44}}{a^5} \left\{ \frac{3}{8} \sin(4\Omega - 4\omega_\oplus t + \delta_{44}) \right. \\ &+ \frac{1}{4} \sin(2M + 2\omega + 4\Omega - 4\omega_\oplus t + \delta_{44}) \end{aligned}$$

$$
\begin{aligned}
&+ \frac{1}{4}\sin(4\Omega - 4\omega_{\oplus}t - 2M - 2\omega + \delta_{44}) \\
&+ \frac{1}{16}\sin(4M + 4\omega + 4\Omega - 4\omega_{\oplus}t + \delta_{44}) \\
&+ \frac{1}{16}\sin(4\Omega - 4\omega_{\oplus}t - 4M - 4\omega + \delta_{44})\Big\}
\end{aligned}
$$

Show that a classical perturbation theory will fail in the vicinity of 12 hour orbits.

Convert V_{44} to the Delaunay elements, and add in the two body Hamiltonian. Discard all but the 2:1 resonance term, and construct a canonical transformation that greatly increases the number of integrals of the motion. How many "equilibrium points" are there near the 2:1 resonance? Compare to the number of GPS satellites in one orbital plane (four).

Problem 2. A small spacecraft is trying to orbit an asteroid which is rotating about its largest (C) principal axis of inertia. The potential function for the spacecraft comes from an earlier chapter

$$
\begin{aligned}
V = &- \frac{\mu}{r} - \frac{\mu}{4r^3}\left\{(3l^2 - 1)(B + C - A)\right. \\
&+ \left.(3m^2 - 1)(A + C - B) + (3n^2 - 1)(A + B - C)\right\}
\end{aligned}
$$

Assume that the spacecraft is within the asteroid's equator plane and is in a nearly circular orbit. Using the principal axis frame, let the asteroid's rotation rate be ω_a.

Expanding to zero order in the eccentricity, show that the direction cosines are

$$
\begin{aligned}
l &\approx \cos(\omega + M - \omega_a t) \\
m &\approx \sin(\omega + M - \omega_a t) \\
n &= 0
\end{aligned}
$$

Write the disturbing function R in terms of the classical orbital elements.

Construct a first order perturbation theory for the semimajor axis a. Show that you have one resonance. At what orbital frequency does this resonance appear?

Recast this problem as a Hamiltonian system using the Delaunay elements. Perform a canonical transformation to a new set of variables adapted to handle the resonance. Physically, what is the resonance angle coordinate? How many equilibrium points are there? Which are stable?

On its fast flyby of the asteroid Ida, the Galileo spacecraft glimpsed a small second body, christened Dactyl, nearly off the long axis of Ida itself. Is this the position of the stable resonant equilibrium?

Problem 3. Expand the equations of motion for the critical inclination, (6.29) and (6.31) about the equilibria, and produce the small deviation linear system that governs their stability. Show that the equilibria on the equator, $g = 0$ and $g = \pi$, are unstable, but that the solutions that put perigee as far north as possible $g = \pi/2$, and as far south as possible $g = 3\pi/2$ are stable.

Chapter 7

Relative Satellite Motion

7.1 Introduction

This chapter brings together some of the numerical techniques available with modern computers, and some of the techniques of perturbation theory. This chapter is essentially the work of the current author.

The problem of building very large antennas in orbit leads to the realization that the antenna does not have to be one single rigid unit. In a *satellite cluster*, the function of the entire phased array antenna is split up among several, possibly many individual satellites. These must be navigated and controlled to within a tenth of the operating wavelength if the swarm is to function as a cohesive single antenna. In the microwave region, the operating wavelength may be in the tens of centimeters, and the dynamics accuracy then has to be around the centimeter level. A related, and far more challenging concept would be an optical phased array antenna. Such telescopes are now common on earth, with a rigid structure behind the individual mirrors, and very precise control. In space it is possible to dispense with the rigid structure, but the mirrors must then be flown to an accuracy of a tenth of a wavelength of light, or a tolerance of several hundred atomic diameters. Achieving this challenge will make it possible to study planets around nearby stars, but it is definitely a challenge.

It is common to use the Clohessy–Wiltshire solution for relative satellite motion to study the motion of a satellite cluster. This solution assumes a circular, two body problem orbit as a reference, and has been extensively used in rendezvous of manned spacecraft. Of course, a shuttle performing a rendezvous with the space station is engaged in a short time, relatively energetic process. The free flying satellites in a cluster must be able to control their relative motion for the useful lifetime of the cluster. In fact, maneuvering will be one factor that limits the lifetime of a cluster.

In this chapter we will replace the circular, two body problem orbit with a periodic orbit about the earth. The linear system solved by Clohessy and

Wiltshire was a constant coefficient system; we will solve a Floquet problem. These two steps will improve the accuracy of the linearized reference solution by about a factor of one thousand compared to using the two body problem. The remaining part of the geopotential will be handled with a perturbation theory based on the Floquet solution, while dissipative forces like air drag may be included by numerically integrating perturbation equations derived from the Floquet solution.

This is possible since modern computer techniques make it possible to reduce the periodic functions of Floquet theory to easily stored Fourier coefficients. The computer then becomes a tool in what amounts to *analytic* work as well, not just "numerical" studies.

7.2 The Reference Periodic Orbit

In dealing with a "real–world" problem, it is almost inevitable that the full system will not be solvable in closed form. It is desirable to include the maximum amount of dynamics within the "solvable" part of the system, since this will give the maximum validity to the unperturbed motion, and reduce to a minimum the dynamics that has to be dealt with via perturbation theory. It will also ensure that any control that must be applied already "knows" about unstable versus benign behavior, and can therefore discriminate between the two. For the earth, a maximal a dynamics model that still permits periodic orbits includes the Newtonian point mass potential and all of the zonal harmonics of the gravity field. This dynamical system is time invariant, so the Hamiltonian is an integral of the motion, and since it is also invariant to rotations about the earth's polar axis, the z component of angular momentum is also constant. This model, of course, includes the very important J_2 zonal harmonic. But it is sometimes claimed that orbital motion about the earth possesses no periodic orbits. The claim that they exist needs to be justified.

At the lowest level, imagine an exactly circular orbit subject to the classic secular perturbations due to J_2. The node will regress, and in a reference frame that moves with the ascending node (which we will call the nodal frame), the circular orbit will close upon itself. Since the orbit is assumed exactly circular, the perigee will advance, but this makes no difference to a circular orbit. Only the orbital period, which is now the period from nodal crossing to nodal crossing, will be changed by the advance of the perigee.

Actually, the statement (5.57) that the argument of perigee must always advance is not true for *very* small eccentricities. The classical secular perturbation theory that leads to this claim begins by ignoring all periodic terms in the disturbing function. These are at least of order one in the eccentricity, and if the eccentricity is very small they should be negligible. The secular part of the disturbing function yields the secular rates, and the effect of the small periodic terms can then be studied. However, the Lagrange planetary equation for the argument of perigee

$$\frac{d\omega}{dt} = \frac{\sqrt{1-e^2}}{na^2 e}\frac{\partial R}{\partial e} - \frac{\cot i}{na^2\sqrt{1-e^2}}\frac{\partial R}{\partial i} \tag{7.1}$$

in terms of the disturbing function R is obviously singular for small eccentricity. Allegedly "negligible" periodic terms in the disturbing function that are of order eccentricity e in the disturbing function are of order $1/e$ in $d\omega/dt$. In other words, they are not negligible at all. The claim that the argument of perigee must always advance is unfounded for nearly circular orbits.

At a much higher level of approximation, there is a considerable literature on "frozen orbits", including many successful applications to actual missions. The theory of frozen orbits treats the behavior of nearly circular orbits under the influence of secular perturbations from all of the zonal harmonics. (Note that we also will isolate all of the zonal harmonics.) By imposing the constraint that the secular rate of the argument of perigee be zero, the theory of frozen orbits produces a very nearly circular orbit which remains nearly "frozen" in its apsidial orientation, while the orbit plane still regresses. These orbits exist for eccentricities so small that the separation of secular and periodic terms is clearly invalid. We will shortly see that the current chapter essentially extends frozen orbit theory by including the periodic terms in the disturbing function, producing a true periodic orbit for the Newtonian plus zonal harmonics gravity problem.

Begin in the usual earth-centered inertial frame, with rectangular coordinates X, Y, Z, with Z along the earth's polar axis. We will use dimensionless units with the radius of the earth $R_\oplus = 1$, the gravitational constant $G = 1$, and the mass of the earth $M_\oplus = 1$. The problem of orbital motion about the earth, restricted to the zonal harmonics, has Hamiltonian function

$$\begin{aligned}\mathcal{H} &= \frac{1}{2}\{P_X^2 + P_Y^2 + P_Z^2\} - \frac{1}{r} \\ &+ \frac{1}{r}\sum_{i=2}^{\infty}\left(\frac{1}{r}\right)^i J_i P_i^0(Z/r),\end{aligned} \tag{7.2}$$

where, of course, $r = \sqrt{X^2+Y^2+Z^2}$, the zonal gravity coefficients $C_{i0} = -J_i$, and $P_i^0(Z/r)$ is the Legendre polynomial of order i. The dynamical system (7.2) possesses time reversal symmetry, and is symmetric about the Z axis. Hence, both the total energy ($\mathcal{H}$) and the Z component of angular momentum

$$L_Z = XP_Y - YP_X \tag{7.3}$$

are conserved. The momenta P_i are just the inertial velocity components, since (7.2) is a Hamiltonian per unit mass of satellite. As a convenient shorthand, we will write the inertial frame state vector as $\mathcal{I}^T = (X, Y, Z, P_X, P_Y, P_Z)$. Note that while the sum over the zonal harmonics will in practice be truncated at some finite limit, in theory *all* of the zonal harmonics can be included in the periodic orbit and relative motion solution.

Nearly circular periodic orbits exist in this system at any inclination and orbital radius, but only in a reference frame regressing about the Z axis at the (as yet unknown) nodal regression rate. For definiteness, pick initial conditions at $t = 0$ on the X axis with a given initial "inclination" i_0 and given initial radius r_0, as

$$\begin{aligned} X &= r_0, \;\; Y = 0, \;\; Z = 0, \\ P_X &= \dot{r}_0, \;\; P_Y = S_0 \cos i_0, \;\; P_Z = S_0 \sin i_0, \end{aligned} \tag{7.4}$$

where the period τ, the radial velocity $\dot{r}_0$, and the tangential speed S_0 are initial parameters to be determined from the solution of a periodic orbit boundary value problem. (The initial radius r_0 and initial inclination i_0 may be specified, since we expect a two parameter family of periodic orbits, as the Hamiltonian has two integrals of the motion.) The three unknown initial conditions are to be determined by enforcing the final boundary conditions at one period, $t = \tau$:

$$\begin{aligned} Z(\tau) &= 0, \\ r(\tau) &= r_0, \\ \mathbf{r}(\tau) \cdot \mathbf{P}(\tau) &= r_0 \dot{r}_0. \end{aligned} \tag{7.5}$$

These conditions state that the satellite must again be crossing the plane $Z = 0$ at τ, that it be at the same distance from the earth r_0, and that it have the same radial velocity $\dot{r}_0$. The radial velocity component $\dot{r}_0$ is necessary since the odd order zonal harmonics (e.g. J_3) introduce a north / south asymmetry into the problem. The particular form of the final conditions (7.5) has been chosen to be independent of the (still unknown) nodal regression rate. Conservation of energy and the Z component of angular momentum then suffice to force both the initial tangential speed S_0 and the initial "inclination" i_0 to return to their initial values.

We have implemented this algorithm numerically. Expanding (7.4) gives the possible changes in the initial conditions as

$$\begin{aligned} \delta \mathcal{I}(0) &= \left\{ \begin{array}{cc} 0 & 0 \\ 0 & 0 \\ 0 & 0 \\ 1 & 0 \\ 0 & \cos i_0 \\ 0 & \sin i_0 \end{array} \right\} \left(\begin{array}{c} \delta \dot{r}_0 \\ \delta S_0 \end{array} \right) \\ &= C \left(\begin{array}{c} \delta \dot{r}_0 \\ \delta S_0 \end{array} \right) \end{aligned} \tag{7.6}$$

Similarly, in the notation of section 1.5 the final condition function can be re-written from (7.5) as

$$\mathbf{G} = \left\{ \begin{array}{c} Z \\ \sqrt{X^2 + Y^2 + Z^2} - r_0 \\ X P_X + Y P_Y + Z P_Z - r_0 \dot{r}_0 \end{array} \right\} = \mathbf{0} \tag{7.7}$$

Linearizing these final conditions, we obtain

$$\delta\mathbf{G} = \frac{\partial\mathbf{G}}{\partial\mathcal{I}(\tau)}\delta\mathcal{I}(\tau) \tag{7.8}$$

$$= \frac{\partial\mathbf{G}}{\partial\mathcal{I}(\tau)}\Phi(\tau,0)\delta\mathcal{I}(0)$$

$$= \frac{\partial\mathbf{G}}{\partial\mathcal{I}(\tau)}\Phi(\tau,0)C\begin{pmatrix}\delta\dot{r}_0\\ \delta S_0\end{pmatrix} \tag{7.9}$$

This supplies the partial derivative matrix of $\mathbf{G}$ with respect to $\dot{r}_0$ and S_0. The third initial parameter is the orbital period, τ. To obtain $\partial\mathbf{G}/\partial\tau$, we need only calculate $d\mathbf{G}/dt$. Then we can set up a Newton–Rhapson algorithm that calculates the changes in the initial values needed to drive $\mathbf{G}\to 0$. Typically, we have observed that all six initial and final conditions match to at least ten and usually twelve significant figures, when compared in the rotating frame of reference.

The orbit does not return to the same position in inertial space. The point at which the orbit closes on itself will have regressed along the equator, and so the orbit is only truly periodic in a rotating frame of reference. Once the periodic orbit has been found, we can calculate the amount of rotation of the initial position about the Z axis $\cos\theta = \mathbf{r}(0)\cdot\mathbf{r}(\tau)/r_0^2$. Then, define an average regression rate for the node as

$$\dot{\Omega} = -\theta/\tau, \tag{7.10}$$

and write a rotation matrix about the Z axis as

$$\mathcal{R}_Z = \left\{\begin{array}{ccc}\cos(\Omega_0+\dot{\Omega}t) & \sin(\Omega_0+\dot{\Omega}t) & 0\\ -\sin(\Omega_0+\dot{\Omega}t) & \cos(\Omega_0+\dot{\Omega}t) & 0\\ 0 & 0 & 1\end{array}\right\}. \tag{7.11}$$

Including the initial right ascension of the node Ω_0 removes any special orientation for the periodic orbit. Then, the rotation transformation

$$\begin{aligned}(X',Y',Z')^T &= \mathcal{R}_Z(X,Y,Z)^T\\ (P'_X,P'_Y,P'_Z)^T &= \mathcal{R}_Z(P_X,P_Y,P_Z)^T\end{aligned} \tag{7.12}$$

is canonical, with a new Hamiltonian

$$\begin{aligned}\mathcal{H}' &= \frac{1}{2}\left\{P_X'^2+P_Y'^2+P_Z'^2\right\}\\ &+ \dot{\Omega}\,(P'_XY'-P'_YX')+V(Z',r)\end{aligned} \tag{7.13}$$

It is in this coordinate frame that the periodic orbit actually closes on itself. We will refer to this reference frame as the nodal frame.

The non-inertial effects of the rotating frame are accounted for by the cross product term $P'_XY'-P'_YX'$ in the Hamiltonian (7.13). The transformation

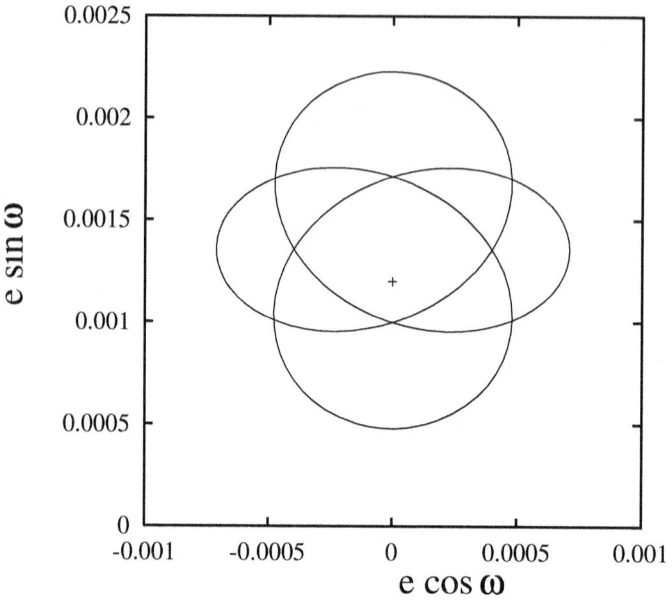

Figure 7.1: The periodic orbit for one revolution plotted on the osculating $e\cos\omega$, $e\sin\omega$ plane.

(7.12) shows that the new momenta P'_i are still the inertial velocity components, now expressed in their components in the rotating nodal frame. This means that non-conservative forces are simply added in the momenta equations of motion, and we need only resolve them in the nodal frame to include their effects. Of course, conservative perturbing forces are included by adding a potential term to the Hamiltonian itself.

Since the system Hamiltonian (7.2) has two integrals of the motion, there is a two parameter family of such orbits, and they exist for any value of the orbital radius and inclination. In fact, the periodic orbit is essentially a J_2 perturbed circular orbit.

The coordinates of the periodic orbit in the nodal reference frame close on themselves, and can be conveniently reduced to Fourier series, see Appendix I. It is this ability to construct a special solution to a dynamical system, and then reduce it to a usable form, that makes the present technique viable. By examining the Fourier coefficients it is very easy to truncate the summation at a level that guarantees a desired accuracy, and of course the series representation can be numerically compared to the periodic orbit to confirm this. We have found that retaining the first 30 terms approaches full double precision accuracy. Writing the canonical state vector in the nodal frame as $\mathcal{N}^T = (X', Y', Z', P'_X, P'_Y, P'_Z)$, the inertial and nodal states are related by $\mathcal{N} = \mathcal{R}_Z^{(2)}\mathcal{I}$, where $\mathcal{R}_Z^{(2)}$ is a six by six block diagonal matrix with two copies of $\mathcal{R}_Z$ on the diagonal. We will write $\mathcal{N}_0(t)$ as a shorthand for the periodic orbit itself.

We return to the relationship between the periodic orbits of the current discussion and frozen orbit theory. Figure 7.1 shows the variation of the osculating elements $e\cos\omega$, $e\sin\omega$ over one cycle of the periodic orbit, for the case of an initial radius of 1.1 earth radii and an inclination of one radian. Also plotted as a point is the approximate values of these quantities in frozen orbit theory, extracted from the plots in Rosborough et al. These were computed for a gravity field including zonal terms through order 50, while we have only used terms through order 14. The agreement is quite satisfactory, given the difference in gravity models and that we are describing slightly different objects. Frozen orbit theory does not include the periodic part of the zonal disturbing potential, while periodic orbit theory includes all of the zonal disturbing potential. The osculating orbital elements show that the argument of perigee ω is librating about $\omega = 90^0$, as is expected in frozen orbit theory. In fact, it cycles *four* times per orbit. Since frozen orbit theory treats only the secular terms in the potential, the current method is a more complete version of that method.

7.3 The Orbital Reference Frame

The reference orbit is periodic in a frame that regresses with the plane of the orbit. But that is not the only frame in which the orbit is periodic. Our goal is to produce a solution that resembles the CW solution, but includes much more of the total dynamics. Also, it is desirable to be able to handle further perturbing forces (e.g.: sectoral and tesseral harmonics) as well as non-conservative forces such as air drag.

Introduce coordinates relative to the periodic orbit

$$\begin{aligned} x &= X' - X_0'(t), \quad p_x = P_X' - P_{X0}'(t) \\ y &= Y' - Y_0'(t), \quad p_y = P_Y' - P_{Y0}'(t) \\ z &= Z' - Z_0'(t), \quad p_z = P_Z' - P_{Z0}'(t) \end{aligned} \tag{7.14}$$

where the quantities subscripted with zero refer to the periodic orbit. This transforms the periodic orbit to the origin of the coordinates. This is canonical, as can be seen by following the method of section 2.12. It reduces the Hamiltonian function to its quadratic and higher order parts, evaluated on the periodic orbit. That is, if we assemble the local coordinates and momenta into a state vector $\mathcal{Y}^T = (x, y, z, p_x, p_y, p_z)$, the new Hamiltonian is

$$\mathcal{H}" = \frac{1}{2!}\mathcal{H}"_{\alpha\beta}\mathcal{Y}_\alpha\mathcal{Y}_\beta + \frac{1}{3!}\mathcal{H}"_{\alpha\beta\gamma}\mathcal{Y}_\alpha\mathcal{Y}_\beta\mathcal{Y}_\gamma + ... \tag{7.15}$$

where repeated Greek indices are summed from one to six, and roman indices (when they occur) can freely take on values from one to six. Here each $\mathcal{H}"_{\alpha\beta...}$

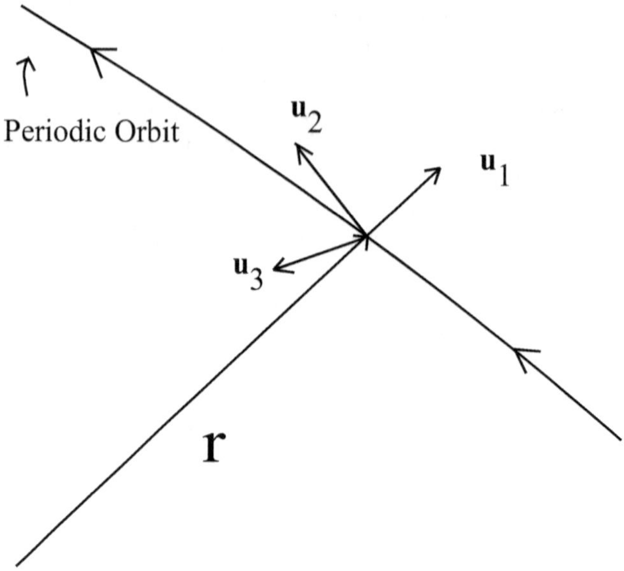

Figure 7.2: The orbital reference frame

is a fully symmetric partial derivative tensor evaluated on the periodic orbit:

$$\mathcal{H}"_{\alpha\beta} = \left.\frac{\partial^2 \mathcal{H}'}{\partial \mathcal{N}_\alpha \partial \mathcal{N}_\beta}\right|_{\mathcal{N}_0(t)}, \tag{7.16}$$

$$\mathcal{H}"_{\alpha\beta\gamma} = \left.\frac{\partial^3 \mathcal{H}'}{\partial \mathcal{N}_\alpha \partial \mathcal{N}_\beta \partial \mathcal{N}_\gamma}\right|_{\mathcal{N}_0(t)}, \tag{7.17}$$

For perturbation and control work, it is necessary to track how to re-introduce any dynamical effects that have been neglected. Forces that we have not yet included in the problem are handled in a very straightforward way. Conservative forces have a potential energy function, and this is expanded about the periodic orbit due to the change of origin (7.14). The zeroth order (in $\mathcal{Y}$) terms perturb the periodic orbit itself, linear and higher terms in $\mathcal{Y}$ perturb the relative motion. Non-potential forces such as air drag are also expanded about the periodic orbit, and since the p_i are still inertial velocity components, their effects appear only in Hamilton's equations for the momenta states. As with conservative forces, only the linear and higher terms in these forces perturb the relative motion.

Now, introduce another set of coordinates that are the standard reference frame used in the Clohessey-Wiltshire solution. As shown in Figure 7.2, define unit vectors tied to the reference periodic orbit, where the first vector is radial, $\mathbf{u}_1 = \mathbf{r}/r$ where $\mathbf{r}$ is the nodal frame position vector, the third vector is the orbit normal $\mathbf{u}_3 = (\mathbf{r} \times \mathbf{v})/|\mathbf{r} \times \mathbf{v}|$ where $\mathbf{v}$ is the nodal frame velocity vector, and the second vector, $\mathbf{u}_2 = \mathbf{u}_3 \times \mathbf{u}_1$, lies near the velocity vector. Assembling these orthogonal unit vectors into a rotation matrix

$$\mathcal{R}^T = \{\mathbf{u}_1\ \mathbf{u}_2\ \mathbf{u}_3\}, \tag{7.18}$$

we can introduce new coordinates in the radial, 'in-track', and orbit normal directions as

$$\begin{pmatrix} q_r \\ q_v \\ q_n \end{pmatrix} = \mathcal{R} \begin{pmatrix} x \\ y \\ z \end{pmatrix}. \tag{7.19}$$

Again, this rotation generates a canonical transformation. The generating function is just

$$F_2 = \begin{pmatrix} p_r \\ p_v \\ p_n \end{pmatrix} \cdot \mathcal{R} \begin{pmatrix} x \\ y \\ z \end{pmatrix}, \tag{7.20}$$

and the momenta transform as $(p_r, p_v, p_n)^T = \mathcal{R}(p_x, p_y, p_z)^T$. Write the new canonical state vector as $\mathcal{Z}^T = (q_r, q_v, q_n, p_r, p_v, p_n)$. It is this set of coordinates that mimic the usual radial / in-track / orbit normal coordinates of the CW solution. (We note here that satellites may use the horizon and orbit normal for their reference frame, rather than the actual radius vector. There is a slight difference due to the non-spherical shape of the earth. This rotation transformation could just as easily use the horizon from an ellipsoidal earth model for the reference direction, and the discussion to follow will be unchanged, although slightly different results will be obtained numerically. The matrix $\mathcal{R}$ need only be periodic to preserve the structure of the Floquet solution to follow.)

The quadratic part of the Hamiltonian (which of course generates a linear dynamical system), transforms into the variables $\mathcal{Z}$ as

$$\begin{aligned} \mathcal{K}_2 &= \frac{1}{2}\mathcal{H}"_{\alpha\beta}\mathcal{Y}_\alpha\mathcal{Y}_\beta + \frac{\partial F_2}{\partial t} \\ &= \frac{1}{2}\mathcal{H}"_{\alpha\beta}\mathcal{R}^{(2)}_{\alpha\gamma}\mathcal{R}^{(2)}_{\beta\delta}\mathcal{Z}_\gamma\mathcal{Z}_\delta + \mathbf{p}\cdot\dot{\mathcal{R}}\mathcal{R}^T\mathbf{q} \end{aligned} \tag{7.21}$$

The matrix $\mathcal{R}^{(2)}$ is just the orbital frame rotation matrix repeated twice

$$\mathcal{R}^{(2)} = \left\{ \begin{array}{cc} \mathcal{R} & 0 \\ 0 & \mathcal{R} \end{array} \right\}, \tag{7.22}$$

and the second term in (7.21) is the cross product to be expected when transforming to a rotating frame of reference. Higher order terms in the Hamiltonian transform more directly, for example the third order contributions to the Hamiltonian (7.21) are

$$\frac{1}{3!}\mathcal{K}_{\alpha\beta\gamma}\mathcal{Z}_\alpha\mathcal{Z}_\beta\mathcal{Z}_\gamma = \frac{1}{3!}\mathcal{H}"_{\lambda\sigma\tau}\mathcal{R}^{(2)}_{\lambda\alpha}\mathcal{R}^{(2)}_{\sigma\beta}\mathcal{R}^{(2)}_{\tau\gamma}\mathcal{Z}_\alpha\mathcal{Z}_\beta\mathcal{Z}_\gamma, \tag{7.23}$$

and so forth. Again, the momenta states p_r, p_v, and p_n are still the inertial velocity components relative to the periodic orbit, but now expressed in the unit vectors of the orbital frame. This means that non-potential forces are simply expanded about the periodic orbit, and then added to Hamilton's equations of motion for the momenta states, once they have been resolved in the orbital reference frame.

7.4 The Floquet Solution

The relative motion solution we are seeking is, like the Clohessy-Wiltshire treatment, a solution to a linear system. However, expanding about the periodic orbit the linearized equations derived from (7.21) are periodic in time. The solution to such problems was first found by Floquet, and has been discussed by many authors since, including the present work, in section 1.7.

The relative motion Hamiltonian (7.21), restricted to its quadratic part, generates the matrix linear system

$$\dot{\Phi} = A\Phi, \quad \Phi(t=0) = I \tag{7.24}$$

where the "plant matrix" A comes directly from Hamilton's equations of motion. Integrating this matrix differential equation for one orbit gives the monodromy matrix $\Phi(\tau, 0)$. Now, according to Floquet, the state transition matrix Φ factors as

$$\Phi(t, 0) = F(t)e^{\mathcal{J}t}F^{-1}(0). \tag{7.25}$$

The periodic modal matrix $F(t)$ and the Jordan form of Poincaré exponents $\mathcal{J}$ must be constructed to find this solution. The initial matrix $F(0)$ is just the eigenvector matrix of $\Phi(\tau, 0)$, as evaluating (7.25) at $t = \tau$ will show. The eigenvalues of the monodromy matrix μ_i, termed the characteristic multipliers, are related to the Poincaré exponents $\mathcal{J}_{ii}$ by $\mathcal{J}_{ii} = \log \mu_i / \tau$. Substituting (7.25) into (7.24) and simplifying produces

$$\dot{F} = AF - F\mathcal{J}. \tag{7.26}$$

This differential equation, propagated for one period, allows the periodic matrix $F(t)$ to be found, and reduced numerically to Fourier series form.

However, the current problem is not quite this simple. Since we have two integrals of the motion, there are two pairs of zero Poincaré exponents, and each has an associated generalized eigenvector. General purpose eigenvalue / vector software is written assuming that the eigenvalues are distinct, and will find only three of the six eigenvectors. But the full set of eigenvectors associated with the four zero Poincaré exponents can be easily constructed from the system's fundamental symmetries, as will now be described. For definiteness, denote the non–degenerate modal vectors as $\mathbf{f}_1$, $\mathbf{f}_2$, with purely imaginary Poincaré exponents. (We are using a numbering scheme that will keep canonically conjugate Floquet modes adjacent to each other in the problem.) The first eigenvector with a zero Poincaré exponent is just the periodic orbit phase space velocity vector, $\mathbf{f}_3 = \mathcal{R}\, d\mathcal{N}_0/dt$ in the orbital reference frame, suitably normalized. The second eigenvector with zero Poincaré exponent is due to symmetry of the problem about the Z axis, and is in the direction of a rigid rotation of the initial conditions about that axis, again expressed in the orbital frame. That is

$$\mathbf{f}_5 = \begin{pmatrix} \mathcal{R}\mathbf{k} \times \mathbf{r} \\ \mathcal{R}\mathbf{k} \times \mathbf{P}' \end{pmatrix} \tag{7.27}$$

where $\mathbf{k}$ is the Z axis unit vector. These are normal eigenvectors of $\Phi(\tau, 0)$. The remaining two eigenvectors are generalized eigenvectors, and satisfy

$$\begin{aligned} (\Phi - I)\,\mathbf{f}_4 &= \tau\mathbf{f}_3, \\ (\Phi - I)\,\mathbf{f}_6 &= \tau\mathbf{f}_5. \end{aligned} \tag{7.28}$$

These equations will still be singular, since $\Phi - I$ has four zero eigenvalues. Note that each generalized eigenvector is undetermined to within addition of an arbitrary multiple of the normal eigenvector: $\mathbf{f}_4' = \mathbf{f}_4 + \gamma\mathbf{f}_3$. Substituting the conditions $\mathbf{f}_j \cdot \mathbf{f}_3 = 0$ and $\mathbf{f}_j \cdot \mathbf{f}_5 = 0$ for two rows of the above will resolve this ambiguity. Also note the extra factor of τ on the right compared to the usual definition of generalized eigenvectors. This factor is necessary to match the Jordan normal form for this type of orbit, which will have two non-zero diagonal elements, and two blocks of the form

$$\left\{ \begin{matrix} 0 & 1 \\ 0 & 0 \end{matrix} \right\}. \tag{7.29}$$

These Jordan blocks give rise to two terms linear in time in the quantity $e^{\mathcal{J}t}$. In turn, this means that there are two directions in phase space where a satellite can drift away from the reference orbit, instead of the usual one direction familiar from the CW solution. The second linear drift direction was first described by Schaub and Alfriend.

The modal matrix $F(t)$ can now be propagated for one orbit via (7.26), and conveniently reduced to Fourier series form. Including the nodal regression rotation, the periodic orbit, and the first order Floquet solution, the inertial position vector of a satellite can be written as

$$\begin{aligned} \mathcal{I}(t) &= R_Z^{(2)\,T}\,\{\mathcal{N}_0(t) \\ &+ (\mathcal{R}^{(2)})^T F(t) e^{\mathcal{J}t} F^{-1}(t_0)\mathcal{Z}(t_0)\Big\}, \end{aligned} \tag{7.30}$$

while in the orbital frame, the position of the satellite relative to the periodic orbit is given by

$$\mathcal{Z}(t) = F(t)e^{\mathcal{J}t}F^{-1}(t_0)\mathcal{Z}(t_0). \tag{7.31}$$

This result includes potentially *all* zonal harmonics of the earth's gravitational field, and like the CW equations is correct to the first order in small displacements from the reference orbit. The expressions (7.30) for the inertial position, and (7.31) for the relative spacecraft positions are conceptually identical to their CW equivalents. However, the current solution includes all zonal harmonic perturbations through the first order in small displacements from the reference orbit, including the very important J_2 term. The solution itself consists of one periodic six vector $\mathcal{N}_0(t)$, expressed as a Fourier series, one periodic six by six matrix $F(t)$, also reduced to Fourier series, the constant matrix $\mathcal{J}$, and the nodal regression rate $\dot{\Omega}$.

Floquet theory has many numerical checks that have been performed in this effort. We have already noted that initial and final conditions in the periodic orbit construction match to essentially double precision accuracy. The degenerate eigenvectors associated with this problem, (7.27) and (7.28), also generally satisfy the eigenvalue problem for Φ to double precision accuracy. A strong check is that the numerical solution to (7.26) is actually periodic, as Floquet theory predicts. We usually achieve this condition to about nine to ten significant figures. Numerically, we observe a very strong decrease in the size of the Fourier coefficients with order, so that truncation becomes a simple matter. Usually about the first 30 Fourier coefficients contain all of the information in the full numerical integration. After reduction of the periodic orbit and modal matrix to truncated Fourier series, we have routinely compared the summed series version to another numerical integration with a different, incommensurable point spacing. This check matches the periodic orbit and modal matrix to well over ten significant figures.

7.5 The Linear Solution Behavior

In this section we describe the Floquet mode shapes of the solution. This will point out the similarities and differences with the more familiar CW solution. All results shown here are for a periodic orbit with an initial $r_0 = 1.1$ earth radii (about 630 km altitude), and an inclination of one radian. We will also show results comparing the CW solution to the current method.

We begin with the one non-degenerate mode, which exists principally in the in-track / vertical direction. It has a non-zero pair of Poincaré exponents which take the form $\mathcal{J}_{11,22} = 0 \pm i\omega_1$, , and the modal frequency ω_1 is of order J_2. Since the periodic modal vectors have a period of one orbit, and the modal frequency is of order J_2, we interpret this mode as the equivalent of the eccentricity mode in the CW solution. The equivalent motion in the CW solution is a simple ellipse in the radial / in-track plane, with a period of one orbit and a ratio of axis lengths of 2:1, and the current mode is very close to this. In the current solution, however, this mode also includes both the secular advance of the perigee and periodic perturbations in the eccentricity itself. The periodic modal vectors $\mathbf{f}_1(t)$ and $\mathbf{f}_2(t)$ (the first two columns of F) model the usual 2:1 ellipse (with some J_2 and higher perturbations), while the Poincaré exponent ω_1 is the expected secular motion of the perigee. The solution (7.31) has been used to construct Figure 7.3, which shows the motion of this mode.

The remaining two Floquet modes are both degenerate, and have one modal

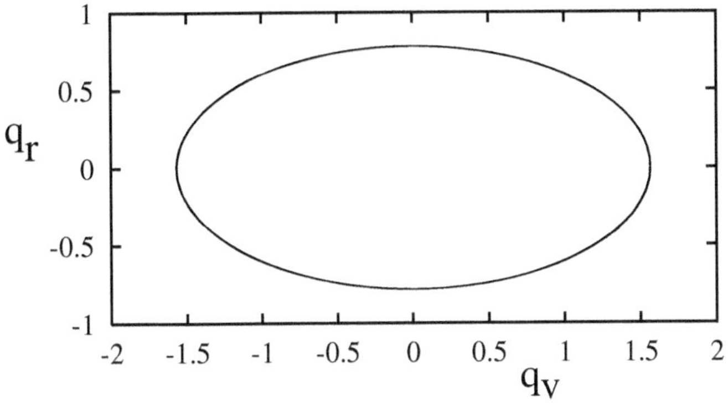

Figure 7.3: The eccentricity mode lies mainly in the vertical / in-track direction.

vector that does not engender a term linear in time in the solution, and another that does. For the degenerate mode caused by the conservation of energy, the modal vector $\mathbf{f}_3$ is the periodic orbit velocity vector. A displacement along this vector will produce a nearly static time displacement between the reference periodic orbit and the satellite. The conjugate extended eigenvector $\mathbf{f}_4$ will produce a linear drift (secular term) in the in-track direction, with a nearly static displacement in altitude. This is a behavior that is conceptually the same as the CW solution, where the corresponding mode combines a change in orbital period with an in-track drift. The modal vector $\mathbf{f}_5$ appears as a nearly "horizontal" oscillation, although not through the origin of the coordinate frame. This mode is a slight change in the node of the nearby satellite relative to the reference periodic orbit. The conjugate extended eigenvector $\mathbf{f}_6$ models the effects of a slight change in inclination (and other elements) which would induce a relative precession of the orbit plane under J_2 perturbations, as found by Schaub and Alfriend. The Floquet solution thus shows all of the behaviors familiar from the CW solution, as well as the expected J_2 effects on the relative motion.

We have also attempted accuracy comparisons between the CW solution and our results. The equations of motion resulting from (7.13) have been integrated numerically using a standard fourth order predictor-corrector integrator which is strongly numerically stable. Initial conditions were chosen using (7.30), so that individual Floquet modes could be selectively studied. Of course, the CW results represent an exact solution within the linear regime for relative motion near a circular orbit in the two body problem, just as our results are an exact solution within the linear regime for motion near a nearly circular periodic orbit in the zonal harmonics problem. Since the Clohessy–Wiltshire solution will be exercised out of its original setting, it will, of course, not do as well as the current method. But the current dynamics model, especially including the first oblateness harmonic J_2, will be much closer to reality than

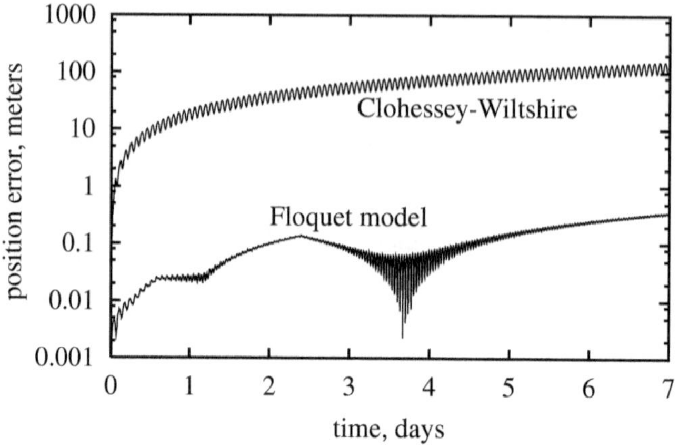

Figure 7.4: The position error for the current (Floquet) solution and the Clohessy-Wiltshire solution.

is the two body problem. We also have elected to give the CW solution every possible advantage, and have only asked it to model the displacements from our precessing periodic orbit, rather than displacements from a non-rotating circular orbit. The CW solution was also given the correct orbital frequency from the periodic orbit. There is not an exact match between the dynamics of (7.13) and our solution either, since higher order nonlinear terms about the reference periodic orbit have been dropped in our method.

Figure 7.4 shows the results of this comparison for a period of one week, with only the eccentricity mode excited. The initial amplitude was approximately 500 meters displacement from the reference orbit, which again was at 630 km altitude with an inclination of one radian. Note that the vertical axis, position error magnitude, is logarithmic, so both solutions are departing roughly linearly in time from the results of the numerical integration. In the case of the Floquet solution, this could either be due to the influence of nonlinear terms which are present in the numerical integration and absent in the linear Floquet solution, or it could be numerical drift in the integration of (7.13). Subtraction of the periodic orbit reference state from the numerical integration produces a large amplification of numerical errors, but the author feels very confident in the accuracy of the numerical integration, and so interprets the discrepancy as the limitations of the first order Floquet theory. The accuracy of the CW solution is uniformly two to three orders of magnitude worse, and within a week the position error reaches the same order of magnitude as the oscillation itself. This occurs in spite of the fact that the CW solution has been given the correct orbital frequency, and has been extended to include the nodal regression of the orbit plane. When used for rendezvous operations, note that the accuracy is fairly good for a day or so, which is usually more than the time interval used in final approach and docking. The "improvement" in the Floquet solution around 3.7 days the author believes to be a unusual situation, where the integration and the theory closely approach each other once per orbit. This, of course, cannot be depended on to always happen.

7.6 Perturbation Theory

The development to here is the analog of Clohessy-Wiltshire theory, with a periodic orbit in a zonal potential replacing the circular two body orbit, and the linear Floquet solution replacing the constant coefficient linear system solution of CW. But this is not yet a complete description of relative satellite motion.

The Floquet problem can be separated in either the nodal or orbital frame of reference. We will find it advantageous to return to the nodal frame Hamiltonian (7.15), since higher order gravity harmonics are more easily expressed in the nodal reference frame. The equations of motion can be written (with non-conservative forces included) as

$$\begin{aligned} \frac{d}{dt}\mathcal{Y}_i &= A_{i\alpha}(t)\mathcal{Y}_\alpha \\ &+ \frac{1}{2!}B_{i\alpha\beta}(t)\mathcal{Y}_\alpha\mathcal{Y}_\beta + \ldots \\ &+ \mathcal{R}_Z^{(2)}\left\{ \begin{array}{c} \mathbf{0} \\ \mathbf{a}_p(t) \end{array} \right\}, \quad i = 1,6 \end{aligned} \tag{7.32}$$

where $\mathcal{R}_Z^{(2)}$ is the rotation matrix from the inertial to the nodal reference frame. Each tensor above is periodic and fully symmetric on every index after the first, and can be written in terms of partial derivatives of the nodal frame Hamiltonian $\mathcal{H}'$:

$$A_{ij}(t) = Z_{i\alpha} \left. \frac{\partial^2 \mathcal{H}'}{\partial \mathcal{N}_\alpha \partial \mathcal{N}_j} \right|_{\mathcal{N}_0(t)}, \tag{7.33}$$

$$B_{ijk}(t) = Z_{i\alpha} \left. \frac{\partial^3 \mathcal{H}'}{\partial \mathcal{N}_\alpha \partial \mathcal{N}_j \partial \mathcal{N}_k} \right|_{\mathcal{N}_0(t)}, \tag{7.34}$$

and so forth. Here Z is the usual symplectic matrix

$$Z = \left\{ \begin{array}{cc} 0 & I \\ -I & 0 \end{array} \right\} \tag{7.35}$$

required to produce the sign structure of Hamilton's equations of motion. As periodic functions, these matrices can be calculated and reduced to Fourier series around the periodic orbit. The nodal frame Hamiltonian $\mathcal{H}'$ includes the two body and zonal harmonics, so all other effects are included in the perturbing acceleration $\mathbf{a}_p$. The quantities $\mathbf{a}_p(t)$ in (7.32) are any extra perturbing acceleration components on the system that were not included in the two body / zonal harmonics baseline model. These are inertial acceleration components transformed to the nodal reference frame by the nodal regression rotation matrix $\mathcal{R}_Z^{(2)}$, and their lowest order terms are those found by evaluation on the periodic orbit itself, with further expansion in $\mathcal{Y}$ components possible.

The Floquet solution makes it possible to separate variables and to introduce new coordinates ideally suited to perturbation theory. Writing the periodic modal matrix in the nodal frame as E, introduce the new variables $\mathbf{z}$ by

$$\mathcal{Y} = E(t)\mathbf{z} = \mathcal{R}^{(2)}(t)F(t)\mathbf{z}. \tag{7.36}$$

That is, $F(t)$ is the orbital reference frame periodic Floquet modal matrix used in previous sections, $\mathcal{R}^{(2)}$ is the six by six orbital to nodal frame rotation matrix, and $E(t)$ is the nodal frame Floquet modal matrix. We will find the use of $E(t)$ more convenient for perturbation work, since higher order gravitational perturbations are more easily expressed in the nodal than in the orbital frame. The very important system state vector $\mathbf{z}(t)$ is the set of six Floquet modal amplitudes. We have used the common transformation of isolating the real and imaginary parts of the one complex mode, so that both $E(t)$ and $F(t)$ in (7.36) are purely real matrices. Then, substituting (7.36) into (7.32), and using the definition of the Floquet modal matrix gives

$$\begin{aligned} \frac{d}{dt}\mathbf{z} &= \left(E^{-1}AE - E^{-1}\dot{E}\right)\mathbf{z} \\ &= \mathcal{J}\mathbf{z}. \end{aligned} \tag{7.37}$$

Here, $\mathcal{J}$ is the constant Jordan normal form for this type of periodic orbit:

$$\mathcal{J} = \left\{ \begin{array}{cccccc} 0 & +\omega_1 & 0 & 0 & 0 & 0 \\ -\omega_1 & 0 & 0 & 0 & 0 & 0 \\ 0 & 0 & 0 & 1 & 0 & 0 \\ 0 & 0 & 0 & 0 & 0 & 0 \\ 0 & 0 & 0 & 0 & 0 & 1 \\ 0 & 0 & 0 & 0 & 0 & 0 \end{array} \right\}. \tag{7.38}$$

where the modal frequency ω_1 is now real. We will refer to the $\mathbf{z}$ as *the modal variables*, and to individual modes as the eccentricity mode (rows and columns 1 and 2, the only mode with purely oscillatory behavior), the energy mode (rows and columns 3 and 4), and the angular momentum mode (rows/columns 5 and 6). Applying this transformation to the entire expansion (7.32) then produces

$$\begin{aligned} \frac{d}{dt}\mathbf{z}_i &= \mathcal{J}_{i\alpha}\mathbf{z}_\alpha \\ &+ \frac{1}{2!}E^{-1}_{i\alpha}B_{\alpha\beta\gamma}E_{\beta\delta}E_{\gamma\epsilon}\mathbf{z}_\delta\mathbf{z}_\epsilon + \ldots \\ &+ E^{-1}\mathcal{R}_Z^{(2)}\left\{ \begin{array}{c} \mathbf{0} \\ \mathbf{a}_p(t) \end{array} \right\}. \end{aligned} \tag{7.39}$$

Again, the term with the perturbing accelerations is only the zeroth order term in an expansion in $\mathbf{z}$. These equations separate variables in the linear part of the system, and are ideal for studying perturbations in the relative motion.

7.7 Two-Body/Zonal Perturbations

The current solution includes the two body problem and all zonal harmonics in both the periodic orbit and the Floquet solution the first order in small quantities with respect to the periodic orbit. One of the largest sources of perturbations then is likely to be the second and higher order terms in the variational equations, especially the two body terms. In the modal variables, the first order solution to (7.39) can be written as

$$\mathbf{z}(t) = e^{\mathcal{J}(t-t_0)}\mathbf{z}(t_0). \tag{7.40}$$

Abbreviating $\delta t = t - t_0$, the matrix $e^{\mathcal{J}\delta t}$ is easily found to be

$$e^{\mathcal{J}\delta t} = \left\{ \begin{array}{cccccc} \cos\omega_1\delta t & \sin\omega_1\delta t & 0 & 0 & 0 & 0 \\ -\sin\omega_1\delta t & \cos\omega_1\delta t & 0 & 0 & 0 & 0 \\ 0 & 0 & 1 & \delta t & 0 & 0 \\ 0 & 0 & 0 & 1 & 0 & 0 \\ 0 & 0 & 0 & 0 & 1 & \delta t \\ 0 & 0 & 0 & 0 & 0 & 1 \end{array} \right\}. \tag{7.41}$$

Of course, the inverse of this matrix is just $e^{-\mathcal{J}\delta t}$.

We begin by assuming that the first order solution (7.40) is sufficiently accurate to be used to evaluate the second order perturbing term. This is the fundamental assumption of perturbation theory. Writing

$$B'_{ijk} = E^{-1}_{i\alpha} B_{\alpha\beta\gamma} E_{\beta j} E_{\gamma k}, \tag{7.42}$$

we note that this is also periodic with the period of the original periodic orbit, and can be developed as a Fourier series. Then, performing the evaluation of the second order terms, (7.39) becomes

$$\frac{d}{dt}\mathbf{z} = \mathcal{J}\mathbf{z} + \mathbf{f}(t), \tag{7.43}$$

where the forcing term is

$$\mathbf{f}_i(t) = \frac{1}{2!} B'_{i\alpha\beta} e^{\mathcal{J}\delta t}_{\alpha\gamma} e^{\mathcal{J}\delta t}_{\beta\delta} \mathbf{z}_\gamma(t_0)\mathbf{z}_\delta(t_0). \tag{7.44}$$

Then, the second order free oscillation solution can be written using standard linear system techniques as

$$\mathbf{z}_i(t) = e^{\mathcal{J}\delta t}_{i\alpha} \left(\mathbf{z}_\alpha(t_0) + \Lambda_{\alpha\beta\gamma}(t)\mathbf{z}_\beta(t_0)\mathbf{z}_\gamma(t_0)\right). \tag{7.45}$$

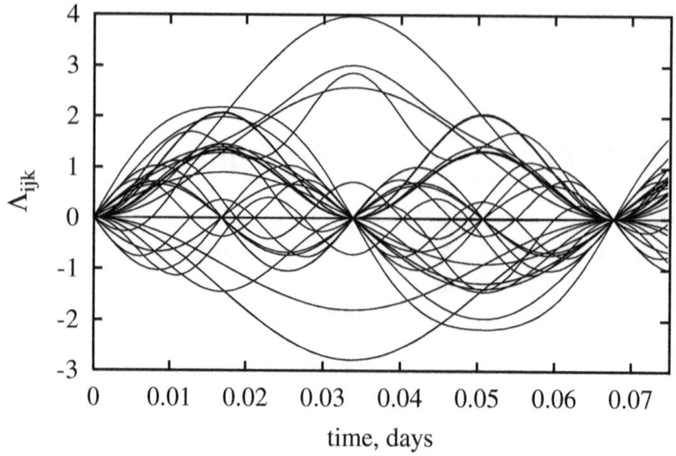

Figure 7.5: Λ_{ijk} components over one orbit, showing only the part with no secular terms.

The first term in the above is the zero order solution (7.40), while the remainder is the desired second order perturbations. The second order perturbations are given by the definite integral

$$\Lambda_{ijk}(t) = \frac{1}{2!}\int_{t_0}^{t} e_{i\alpha}^{-\mathcal{J}\delta t} B'_{\alpha\beta\gamma} e_{\beta j}^{\mathcal{J}\delta t} e_{\gamma k}^{\mathcal{J}\delta t} dt. \tag{7.46}$$

Closed form evaluation of this integral is detailed in Appendix II.

Examining (7.46), and remembering that B' is periodic, we might expect secular terms (terms growing linearly in time) and mixed secular terms (terms involving both a periodic term and a term proportional to time) for those components of Λ_{ijk} involving a term in time in (7.41). Secular terms could also arise from a non-zero constant term in the B' Fourier series. The existence of further secular and mixed secular terms in the second order solution is important, since it governs whether or not more directions of instability appear, other than the two directions present in the first order solution. Figure 7.5 shows the behavior of the Λ_{ijk} over slightly more than one orbit. Components of Λ_{ijk} for $i \neq 3, 5$, and $j, k \neq 4, 6$ have been suppressed. That is, we have excluded the two directions z_3 and z_5 already known to possess secular terms, and excluded the two initial conditions $z_4(t_0)$ and $z_6(t_0)$ known to cause such secular terms. The resulting figure shows that the remainder of Λ_{ijk} is periodic. This means that no new unstable behavior appears at the second order for zonal perturbations, other than that which appeared in the first order solution.

However, this does not mean that if the modal variables $z_4 = 0$ and $z_6 = 0$ that the cluster will remain together. There *are* secular terms due to nonzero values of $z_i, i \neq 4, 6$ that produce, in particular, in-track dispersion of the cluster. These must be calculated to accurately keep the cluster together in one region of space.

7.8 Perturbations from Outside Forces

In order to construct a periodic orbit, some forces were excluded from the baseline dynamics. While the inclusion of the zonal harmonics brings the important J_2 term into the reference solution, all sectoral and tesseral gravitational harmonics have been excluded, as well as air drag. There is a fairly simple way that their effects may be included in the solution, however. Since they have been excluded from the baseline dynamics and the Taylor's series expansion of the dynamics about the periodic orbit, they first appear in the nodal frame variational equations (7.32) or the modal variational equations (7.39) with lowest order terms that are the force *evaluated on the periodic orbit itself.* The next order terms will involve the differential effects of these forces across the span of the satellite cluster: the first order terms in $\delta\mathcal{N}$ or $\mathbf{z}$. These higher order terms will involve, for example, the differential gravitational acceleration of a tesseral harmonic term across the diameter of the cluster, or the extra air drag experienced by an individual satellite because of its motion about the cluster center. The effects of these first order perturbations will be extremely small compared to the already small effects of the zero order terms.

Since both the nodal frame system (7.32) or the modal frame system (7.39) are linear systems, and since these extra forces, when evaluated on the periodic orbit become functions of time alone, we obtain a linear system with a time dependent forcing function. One simple way to incorporate these forces is to obtain the particular solution to the linear system by numerical integration. The nodal variable particular solution is the result of integrating

$$\frac{d}{dt}\mathcal{Y}_i = A_{i\alpha}\mathcal{Y}_\alpha + \mathcal{R}_Z^{(2)}\left\{\begin{array}{c}\mathbf{0}\\ \mathbf{a}_p\end{array}\right\}, \tag{7.47}$$

with $\mathcal{Y}_i = 0$ at the initial time. This form can be thought of as including the perturbations from outside forces in the description of the periodic orbit itself. Alternately, (and to this author preferably) the modal frame particular solution is the result of numerically integrating

$$\frac{d}{dt}\mathbf{z}_i = \mathcal{J}_{i\alpha}\mathbf{z}_\alpha + E^{-1}(t)\mathcal{R}_Z^{(2)}\left\{\begin{array}{c}\mathbf{0}\\ \mathbf{a}_p\end{array}\right\}, \tag{7.48}$$

again with zero initial conditions $\mathbf{z}_i = 0$ at the initial time. Of course, the purpose in calculating the particular solution is so that the effects of these perturbations can be subtracted to give the "free" oscillations. This means that the satellite's control system will not be burdened with maneuvers to null out benign perturbations of the cluster as a whole.

While we are numerically integrating (7.48) to obtain the forced perturbations, our method still holds a great advantage over a simple all-encompassing numerical integration of the equations of motion in, say, the inertial frame. Such a numerical integration would not give the analyst any insight into which part of the integration produces effects that can be tolerated (e.g. purely periodic effects), and which part of the integration will tear the cluster apart. A modern satellite theory must inevitably resort to numerical integration to handle the wide variety of forces that must be included in a state of the art model. But the current theory identifies and separates the acceptable and unacceptable effects to a very high order of accuracy.

One important modification can be made to these equations that greatly extends their range of validity. The periodic orbit and $E(t)$ matrix are represented by Fourier series, and it is a simple matter to insert an arbitrary phase angle in the evaluation of these series, $\omega_0 t \to \omega_0 t + \phi$, where $\omega_0 = 2\pi/\tau$ is the periodic orbit frequency. Now, a small change in this phase angle will cause a small displacement along the state space velocity vector. But this is redundant with the third Floquet mode z_3, which is also a displacement along the state space velocity vector. Using a phase angle in the periodic orbit is a global description of this effect, while the Floquet mode z_3 is the equivalent local description. Since we expect significant in-track perturbations, especially due to air drag, we have elected to replace the local in-track mode z_3 with the global phase angle ϕ in the periodic orbit. The equivalent in-track displacements are

$$\dot{\mathbf{X}}_0 \delta t = \dot{\mathbf{X}}_0 \delta\phi/\omega_0 = \mathbf{f}_3 z_3. \tag{7.49}$$

From this we find

$$\phi = \omega_0 \frac{|\mathbf{f}_3|}{\left|\dot{\mathbf{X}}_0\right|} z_3, \tag{7.50}$$

using the fact that $\mathbf{f}_3$ and $\dot{\mathbf{X}}_0$ are strictly colinear. Then, reducing this to a differential equation gives

$$\dot{\phi} = \omega_0 \frac{|\mathbf{f}_3|}{\left|\dot{\mathbf{X}}_0\right|} \dot{z}_3, \tag{7.51}$$

which will completely replace the $\dot{z}_3$ equation of motion in (7.48). Since we are effectively declaring $z_3 \equiv 0$ for all time, there is no term in the above involving time derivatives of the vector magnitudes.

7.9 Two–Impulse Control

In general, with only the ability to impulsively maneuver a satellite, it will take two maneuvers to reposition a satellite from an initial state $\mathbf{z}(t_0)$ to a given final state $\mathbf{z}(t_3)$. We will almost always wish to specify final modal amplitudes $z_4 = 0$ and $z_6 = 0$ to prevent drift from the cluster reference orbit. (The one exception is initial deployment.) Imposing these two conditions places too many constraints on one maneuver, while they can be handled as a matter of course with two maneuvers. We will think of these maneuvers occurring at times t_1

and t_2, with the maneuver times imbedded within the overall time window as $t_0 \le t_1 < t_2 \le t_3$. The overall time interval $(t_0,\ t_3)$ can be interpreted as an operational constraint: an interval of time when the cluster will not be needed for its primary mission, and orbit adjustments can be made. But we shall also see that the introduction of bounding times t_0 and t_3 also has the effect of making the maneuver optimization time-independent.

The modal state propagates with time as

$$\mathbf{z}(t) = e^{J(t-t_0)}\mathbf{z}(t_0), \tag{7.52}$$

and the modal and orbital frame variables are related as

$$\mathbf{X}(t) = \left\{ \begin{array}{c} \mathbf{r}(t) \\ \mathbf{v}(t) \end{array} \right\} = F(t)\mathbf{z}(t). \tag{7.53}$$

We recall from the end of section 7.3 that the orbital frame velocity $\mathbf{v}$ is actually the inertial velocity of the spacecraft, but resolved along the orbital frame axis system. At one level this is an artifact of Hamiltonian dynamics, but it is very welcome here, since maneuvers appear to directly change the inertial velocity. In an impulsive maneuver, then, where

$$\begin{aligned} \mathbf{r} &\rightarrow \mathbf{r} \\ \mathbf{v} &\rightarrow \mathbf{v} + \Delta\mathbf{v}, \end{aligned} \tag{7.54}$$

the modal variables undergo the change

$$\mathbf{z} \rightarrow \mathbf{z} + F^{-1}(t) \left\{ \begin{array}{c} \mathbf{0} \\ \Delta\mathbf{v} \end{array} \right\}. \tag{7.55}$$

So, beginning at the initial modal state $\mathbf{z}(t_0)$, we propagate forward to the first maneuver time t_1, and there perform the first maneuver. The modal state just after this maneuver is given by

$$\mathbf{z}(t_1^+) = e^{J(t_1-t_0)}\mathbf{z}(t_0) + F^{-1}(t_1) \left\{ \begin{array}{c} \mathbf{0} \\ \Delta\mathbf{v}_1 \end{array} \right\} \tag{7.56}$$

Then propagating the above result to the time of the second maneuver, t_2, and performing this maneuver gives

$$\begin{aligned} \mathbf{z}(t_2^+) &= e^{J(t_2-t_0)}\mathbf{z}(t_0) + e^{J(t_2-t_1)}F^{-1}(t_1) \left\{ \begin{array}{c} \mathbf{0} \\ \Delta\mathbf{v}_1 \end{array} \right\} \\ &+ F^{-1}(t_2) \left\{ \begin{array}{c} \mathbf{0} \\ \Delta\mathbf{v}_2 \end{array} \right\} \end{aligned} \tag{7.57}$$

Finally, we propagate to the standard final time t_3, chosen to be the desired state at a time larger than the time interval we expect to search for optimal maneuvers. The modal state then is

$$\begin{aligned} \mathbf{z}(t_3) &= e^{J(t_3-t_o)}\mathbf{z}(t_o) \\ &+ e^{J(t_3-t_1)}F^{-1}(t_1)\left\{ \begin{array}{c} \mathbf{0} \\ \Delta\mathbf{v}_1 \end{array} \right\} \\ &+ e^{J(t_3-t_2)}F^{-1}(t_2)\left\{ \begin{array}{c} \mathbf{0} \\ \Delta\mathbf{v}_2 \end{array} \right\}. \end{aligned} \tag{7.58}$$

This must equal our specified state at the final time.

Then with some manipulation, (7.58) can be put into the form

$$\Gamma(t_2, t_1)\Delta\mathbf{V} = \mathbf{b}. \tag{7.59}$$

The vector on the right side,

$$\mathbf{b} = \mathbf{z}(t_3) - e^{J(t_3-t_o)}\mathbf{z}(t_o) \tag{7.60}$$

does not depend on either maneuver time, but instead depends only on the difference between the chosen final state $\mathbf{z}(t_3)$ and the initial conditions naturally propagated to the same final time t_3. The unknown vector

$$\Delta\mathbf{V} = \left\{ \begin{array}{c} \Delta\mathbf{v}_1 \\ \Delta\mathbf{v}_2 \end{array} \right\} \tag{7.61}$$

includes both impulsive maneuvers in one vector. All maneuver-time dependence appears in the matrix Γ, which has six rows and columns. Its first three columns are columns 4 through 6 of the matrix $\exp(J(t_3 - t_1))F^{-1}(t_1)$, while the second three columns of Γ are columns 4 through 6 of the matrix $\exp(J(t_3 - t_2))F^{-1}(t_2)$. While this appears to depend on the final time t_3 also, in fact the modal variables are nearly constant on the timescale of one orbit, which is the major application that we have in mind. In this case, $\mathbf{b}$ will be a small vector. For initial deployment of the satellite cluster, it might be desirable to allow deployment to occur over several orbits. Also, notice immediately that Γ is singular when $t_2 - t_1$ is a multiple of the period τ of the orbit, since the matrix $F(t)$ is periodic with this period.

The introduction of the final time t_3 is necessary. Some possible maneuvers, for example adjusting a satellite's relative inclination, are both expensive and do not become less expensive with time. Other possible maneuvers, for example adjusting the along-track displacement of a satellite, become less expensive the greater the time difference $t_2 - t_1$, and their cost theoretically goes to zero as $t_3 \to \infty$. This is unrealistic in the stationkeeping application, where the satellites must be returned to their nominal trajectories at infrequent intervals, although a larger value of t_3 might be desirable for initial deployment. In this work we will choose $t_3 - t_o$ to be about one and one-half orbits, in order to not overly constrain the maneuver optimization, while at the same time not allowing the maneuvers to stretch over too long a period of time. After maneuvering, each satellite will have to determine its precise trajectory again, and depending on the cluster's mission this may have to happen before normal operations can be fully resumed. Hence, we have elected to keep the interval over which maneuvers are to be performed relatively short.

Some attention must also be given to the choice of how the definition of "optimal" will be implemented. Our goal will be to minimize fuel usage in the long term. If the spacecraft is capable of full reorientation before each maneuver, then maneuvers are done with no geometric wastage. Then an appropriate cost function to minimize is

$$\mathcal{C}_I = |\Delta \mathbf{v}_1| + |\Delta \mathbf{v}_2| \tag{7.62}$$

Alternately, (and far more probably for a small spacecraft), the vehicle may not be free to completely reorient itself for maneuvers. If it is horizon-stabilized in the orbital reference frame but still free to roll about the vertical axis, then horizontal and vertical components of the maneuver are done separately, and an appropriate cost function is

$$\mathcal{C}_{II} = \sum_{i=1}^{2} \sqrt{\Delta v_{i1}^2 + \Delta v_{i2}^2} + \sqrt{\Delta v_{i3}^2} \tag{7.63}$$

Finally, a fully horizon-stabilized satellite might also keep preferential axes aligned with the in-track and orbit normal directions, and therefore would essentially do separate maneuvers in all three coordinate axes. Then an appropriate maneuver cost function will be

$$\mathcal{C}_{III} = \sum_{i=1}^{2} \sum_{j=1}^{3} \sqrt{\Delta v_{ij}^2} \tag{7.64}$$

Figure 7.6 shows a typical stationkeeping maneuver optimization. About one orbital period is shown in both t_1 and $t_2 - t_1$ along the time axes. Large "ridges" appear in the cost near both one orbital period $\tau \approx 100$ minutes, and at approximately $\tau/2$ in $t_2 - t_1$. The singularity near $\tau/2$ is also expected, since the modal matrix $F(t)$ is periodic with one period, and so near $\tau/2$ some column vectors in F become nearly the negatives of their initial directions. The diagonal edge is due to the constraint that $t_2 \leq t_3$, which here was set slightly over one period. As we allow the final time t_3 to grow, some maneuvers (e.g. in-track displacements, z_3) will become cheaper, while orbit plane changes z_5 and free eccentricity / argument of perigee changes z_1, z_2 will cost essentially the same amount. In fact, if the exponential matrix factors imbedded within $\Gamma(t_2, t_1)$ were constant, then Γ would be doubly periodic in both t_1 and $t_2 - t_1$, accounting for much of the appearance of Figure 7.6. The optimal two-impulse burn to reposition a satellite within the cluster can then be found by using any reliable global minimization software on the system (7.59).

The case shown in Fig 7.6 is one that arose in the simulations. Minimum maneuver cost in this case is less than 0.3 cm/sec. Again, prominent ridges near the orbital period and half the period appear. Also, while this figure shows two

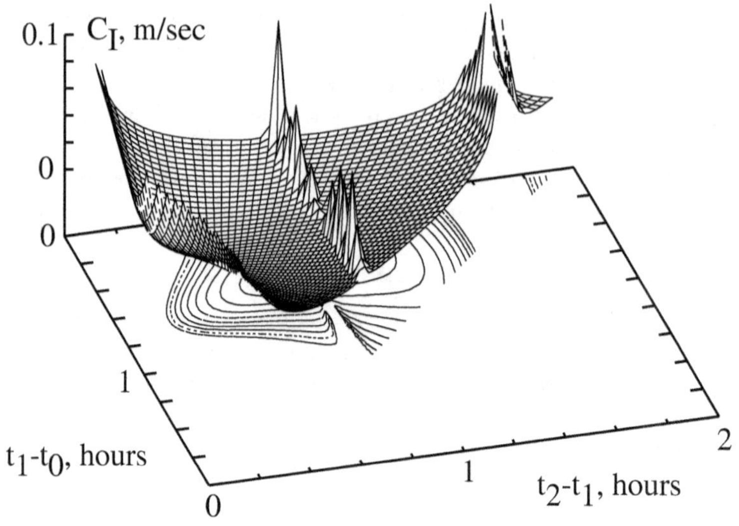

Figure 7.6: Maneuver cost C_I for a typical stationkeeping two impulse maneuver.

minima interior to the region searched, the actual minima may also appear at a corner or along the edge. The constraint boundary $t_2 > t_1$ is a perfectly legitimate constraint, but the choice of the end of the maneuver window final time t_3 we see as more in the nature of an operational constraint. Perhaps a satellite cluster whose principal mission is to observe earth's land areas might use several consecutive orbits lying mostly over the oceans to perform the stationkeeping maneuvers, and that this would dictate the stationkeeping time interval (t_0, t_3).

7.10 References

This chapter is based on a series of papers by the author, which may be found in the Journal of Guidance, Control and Dynamics, specifically

All discussions of relative satellite motion refer back to Clohessy and Wiltshire, who described the solution that is still used for rendezvous today. It is covered in many introductory texts, for example Wiesel, "Spaceflight Dynamics", section 3.5.

For "frozen" orbits, among many worthwhile efforts, we note Nickerson et al. (Journal of the Astronautical Sciences volume XXVI, 1978 (for an early discussion, while Rosborough and Ocampo (Advances in the Astronautical Sciences volume 76, 1991) contains a more through treatment, and Shapiro (Advances in the Astronautical Sciences, volume 99, 1991) discusses flight experience with

Topex/Posedion and contains a through bibliography.

The second linear drift direction was first described by Schaub and Alfriend. It corresponds, of course, to differential nodal regression in J_2 perturbed orbits.

While extended eigenvectors are covered in any good linear analysis book, it is rarely necessary to calculate them in practice, since with a numerical matrix one cannot really be sure of a repeated eigenvalue. An exception is Hamiltonian mechanics, where each integral of the motion assures us of a pair of zero Poincaré exponents. More details on the construction of degenerate Floquet modal eigenvectors can be found in Wiesel (Celestial Mechanics, volume 23, 1981) and Wiesel and Pohlen (Celestial Mechanics and Dynamical Astronomy, volume 58, 1994).

7.11 Appendix I

The reduction of a periodic orbit and its modal vectors into Fourier series is the key to actually using them in further developments. This algorithm is given in Brouwer and Clemence, "Methods of Celestial Mechanics", Academic Press, New York, 1961, pages 108–113, and is repeated here for completeness. Any periodic function can be expressed as a Fourier series

$$F(\theta) = \frac{1}{2}c_0 + \sum_{i=k}^{\infty} \{c_k \cos k\theta + s_k \sin k\theta\} \tag{7.65}$$

Given a table of $2N$ evenly spaced samples of $F(\theta)$, where the values start at $\theta = 0$ and would end with $\theta_{2N+1} = 2\pi$, then each sample generates one linear equation (7.65) in the unknown coefficients c_k, s_k. The result of solving for the coefficients gives

$$c_k = \frac{1}{N}\sum_{j=0}^{2N-1} F(j\alpha)\cos kj\alpha \tag{7.66}$$

$$s_k = \frac{1}{N}\sum_{j=0}^{2N-1} F(j\alpha)\sin kj\alpha \tag{7.67}$$

where $\alpha = \pi/N$. These apply for $k = 0, 1, ...n-1$ in both cases, although there is no s_0 coefficient. The higher order c_k, $s_k \to 0$ if the Fourier series is to be an accurate representation of the periodic function.

7.12 Appendix II

The second order terms in the zonal problem require the evaluation of the integral (7.46). We wish to perform this task analytically to see if any further instabilities appear. The state transition matrix for the first order solution is given by (7.41), and the tensor B'_{ijk} from (7.42) is available as a Fourier series. Counting up the possibilities from $\exp(\mathcal{J}\delta t)$, we can have terms involving $\cos\omega_1\delta t$, $\sin\omega_1\delta t$, and t, raised to powers up to three. There will also be contributions from the Fourier series representation of the B' tensor in sines and cosines of $\omega_0 t$, where ω_0 is the frequency of the periodic orbit. Then suppressing all i, j, and k dependence in (7.46), we must integrate objects of the form

$$\begin{aligned}\Lambda &= \int_{t_0}^{t}\sum_{\alpha\beta\gamma}^{6}\sum_{\epsilon=0}^{\infty}\frac{1}{2}(\pm 1)\left\{c_\epsilon\cos\epsilon\omega_0 t + s_\epsilon\sin\epsilon\omega_0 t\right\}\\ &\times\ (\cos\omega_1 t)^{n_1}(\sin\omega_1 t)^{n_2}\, t^{n_3}dt. \end{aligned} \tag{7.68}$$

Here c_ϵ and s_ϵ are Fourier series coefficients from a particular term of $B'_{\alpha\beta\gamma}$, and the integers n_1, n_2, and n_3 come from counting the occurrence of these terms in the particular term of the tensor product in (7.46), as does the factor ± 1.

Now, we replace the trigonometric functions with their complex exponential equivalents. The B' Fourier series term becomes

$$c_\epsilon\cos\epsilon\omega_0 t + s_\epsilon\sin\epsilon\omega_0 t = \mathcal{C}e^{i\epsilon\omega_0 t} + \mathcal{C}^* e^{-i\epsilon\omega_0 t} \tag{7.69}$$

where $\mathcal{C} = (c_\epsilon - is_\epsilon)/2$, and $*$ denotes the complex conjugate. The standard identities are used to replace $\cos\omega_1 t$ and $\sin\omega_1 t$ with their complex equivalents. Then, using the binomial theorem twice gives

$$\begin{aligned}\Lambda &= \int_{t_0}^{t}\sum_{\alpha\beta\gamma}^{6}\sum_{\epsilon=0}^{\infty}\left(\frac{1}{2}\right)^{1+n_1+n_2}(\pm 1)(-i)^{n_2}\\ &\quad \sum_{\lambda=0}^{n_1}\sum_{\sigma=0}^{n_2}\frac{n_1!}{(n_1-\lambda)!\lambda!}\frac{n_2!(-1)^{n_2}}{(n_2-\sigma)!\sigma!}\\ &\times\ \int_{t_0}^{t}\left\{\mathcal{C}e^{\upsilon_1 t}t^{n_3} + \mathcal{C}^* e^{\upsilon_2 t}t^{n_3}\right\}dt. \end{aligned} \tag{7.70}$$

Here,

$$\begin{aligned}\upsilon_1 &= i(\epsilon\omega_0 + (n_1+n_2-2\lambda-2\sigma)\omega_1,\\ \upsilon_2 &= i(-\epsilon\omega_0 + (n_1+n_2-2\lambda-2\sigma)\omega_1, \end{aligned} \tag{7.71}$$

and $0! = 1$ when it occurs.

The actual integrations have now been reduced to elementary integrals. For completeness we cite the necessary results

$$\begin{aligned}\int e^{\upsilon t}dt &= \frac{1}{\upsilon}e^{\upsilon t},\\ \int te^{\upsilon t}dt &= \frac{\upsilon t-1}{\upsilon^2}e^{\upsilon t}, \end{aligned}$$

$$\begin{aligned}\int t^2 e^{\upsilon t} dt &= \frac{\upsilon^2 t^2 - 2\upsilon t + 2}{\upsilon^3} e^{\upsilon t}, \\ \int t^3 e^{\upsilon t} dt &= \frac{\upsilon^3 t^3 - 3\upsilon^2 t^2 + 6\upsilon t - 6}{\upsilon^4} e^{\upsilon t}.\end{aligned}$$

These integrals will be real valued, although they are most conveniently evaluated with complex arithmetic. Also, for $\epsilon = 0$ there are some values of the summations over λ and σ which lead to one or both of the $\upsilon_i = 0$. In this case, the integrals become the familiar

$$\int t^{n_3} dt = \frac{1}{n_3 + 1} t^{n_3+1}. \tag{7.72}$$

It is the terms containing powers of time that are of concern in the free second order solution, the secular and mixed secular terms. We have taken the trouble to construct this semi-analytical solution (using numerically determined Fourier coefficients for B') to the second order perturbation problem in order to determine where such terms occur. In practice it may be simpler to include the second order two body / zonal perturbations by direct numerical integration of (7.39) instead of (7.48).

Chapter 8

KAM Tori

8.1 Introduction

The KAM theorem, named after Kolmogorov, Arnol'd, and Moser, is a critical result dating from the 1950's. It states that for weak perturbations on a Hamiltonian system in the absence of resonance, that most of the solutions of the unperturbed system persist, and are only slightly changed. Since the unperturbed system is presumed to have periodic dependence on one or more angle variables in a Hamilton-Jacobi solution, the perturbed structures are tori: they are topologically the equivalent of "donut" shapes in the phase space.

After a very long time when the KAM theorem was the province of pure mathematicians, there is a resurgence of interest in actually constructing such objects. In this chapter we will examine the spectral analysis method, mainly due to Laskar, with some applications to earth orbits, principally by the author and his students.

Since a KAM torus is a geometric structure, periodic in each direction around the surface, it must be possible to represent the "physical" coordinates $\mathbf{q}$ as a Fourier series

$$\mathbf{q} = \sum_{\mathbf{j}} (\mathcal{C}_{\mathbf{j}} \cos \mathbf{j} \cdot \mathbf{Q} + \mathcal{S}_{\mathbf{j}} \sin \mathbf{j} \cdot \mathbf{Q}) \tag{8.1}$$

This is a multiple Fourier series, with vector coefficients $\mathcal{C}_{\mathbf{j}}$, $\mathcal{S}_{\mathbf{j}}$, a vector summation index $\mathbf{j}$, and a vector of new coordinates $\mathbf{Q}$. The dot products $\mathbf{j} \cdot \mathbf{Q}$ are the linear combinations of angles we have come to expect in perturbation theory work. Furthermore, the new coordinates Q_i increment linearly with time

$$Q_i(t) = \omega_i t + Q_{i0} \tag{8.2}$$

so there is a set of *basis frequencies* ω_i which are constant across the surface of the torus. This is literally the action-angle solution promised by Hamilton-Jacobi theory.

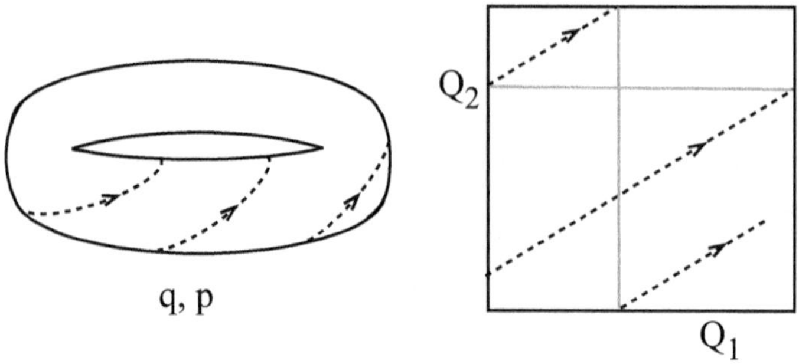

Figure 8.1: A KAM torus in physical coordinate q, p space, and in the new coordinates Q_i.

[Some additional comments are in order about the summation indices j_i. Since sines obey $\sin(-x) = -\sin x$, and cosines obey $\cos(-x) = \cos x$, some combinations of the $\mathbf{j}$ are redundant. To avoid singularities when trying to determine coefficients, the first non-zero j_i, in order, should be greater than zero.]

Figure 8.1 shows a schematic KAM torus on the left, and the new coordinate space on the right. This is a notional representation, since a minimal KAM torus would have two new coordinates Q_i, but would be imbedded in a four dimensional space of the old coordinates and momenta. There would be four phase space variables, q_1, q_2, p_1, p_2, so a number of dimensions have been suppressed in the figure. Also, as the Hamilton-Jacobi solution, the new coordinates Q_i span only half the dimension of the phase space. The new momenta must be the remaining dimensions of the phase space, and represent directions off the surface of the torus. While the new momenta $\mathbf{P}$ do not appear explicitly in (8.1), their numerical values can be calculated from the Poincaré integral invariants. These state that

$$\frac{1}{2\pi}\oint_{\Gamma} \mathbf{P}\cdot d\mathbf{Q} = \frac{1}{2\pi}\oint_{\Gamma} \mathbf{p}\cdot d\mathbf{q} \tag{8.3}$$

where Γ is any arbitrary closed contour in phase space. But if we pick the contour Γ_i to be a line where only one coordinate Q_i is changing and all the other Q_j are constant, then the left integral becomes

$$\frac{1}{2\pi}\oint_{\Gamma_i} \mathbf{P}\cdot d\mathbf{Q} = \frac{1}{2\pi}\int_0^{2\pi} P_i dQ_i = P_i \tag{8.4}$$

The torus expression (8.1) enables us to transform the right side integral as

$$P_i = \frac{1}{2\pi}\oint_{\Gamma_i} \mathbf{p}\cdot d\mathbf{q} = \frac{1}{2\pi}\int_0^{2\pi} \mathbf{p}\cdot\frac{\partial \mathbf{q}}{\partial Q_i} dQ_i \tag{8.5}$$

and this last form can be integrated by numerical quadratures. The fact that this is a Poincaré integral *invariant* means that which contour Γ_i we choose [e.g. the constant values of the other Q_j while allowing only one Q_i to vary] is irrelevant.

The KAM theorem does not pertain near a resonance of the unperturbed system. In that case, we cannot use the KAM theorem to claim that tori exist. However, since we already know that nonlinear resonance can produce significant perturbations from the unperturbed system, this should not be surprising. The existence of chain of island structures on surfaces of section would seem to indicate that tori can indeed exist near resonance. Resonant tori simply will not represent small perturbations from the unperturbed system.

In this chapter the main application will be earth satellite orbits dictated by the earth's geopotential. KAM tori about the earth will be static, geometric structures in the earth centered rotating frame, where the earth's gravity field is a static and geometric structure. The Hamiltonian is then

$$\begin{aligned} \mathcal{H} &= \frac{1}{2}\left(p_x^2 + p_y^2 + p_z^2\right) + \omega_\oplus \left(y p_x - x p_y\right) \\ &- \frac{\mu}{r} \sum_{n=1}^{\infty} \sum_{m=1}^{n} \left(\frac{r}{R_\oplus}\right)^{-n} P_n^m \left(\sin\delta\right) \\ &\times \left(C_{nm}\cos m\lambda + S_{nm}\sin m\lambda\right) \end{aligned} \tag{8.6}$$

Here μ is the gravitational parameter, $R_\oplus$ is the equatorial radius of the earth, and C_{nm}, S_{nm} are the field coefficients that complete the gravity model. The functions P_n^m are the associated Legendre polynomials, and the radius r, geocentric latitude δ, and east longitude λ are found from

$$\begin{aligned} r &= \sqrt{x^2 + y^2 + z^2} \\ \sin\delta &= \frac{z}{\sqrt{x^2+y^2}} \\ \tan\lambda &= \frac{y}{x} \end{aligned} \tag{8.7}$$

Since there is no actual intermediate solvable system, there is no real benefit to posing this problem in, for example, the Delaunay elements.

8.2 The Finite Fourier Transform

The spectral analysis method, to which Jacques Laskar is the most notable contributor, is currently the preferred method to construct a KAM torus. A long numerical integration of the dynamical system in question is subjected to a Fourier transform, but of a rather careful nature. The Fast Fourier transform is justifiably famous, for it optimizes the execution speed for processing a batch of data. It can very quickly produce overall spectra of a signal, here the numerically integrated coordinate history $\mathbf{q}(t)$. But it derives its speed from clever use of sample spacing, usually restricted to a power of two, for efficiency reasons. Alternately, if the integrated orbit really is a Fourier series, then the Fourier transform as the total integration time $T \to \infty$ would become a series of delta functions: infinitely tall spikes with infinitesimal width. These would be very difficult to deal with, but allowing $T \to \infty$ is out of the question anyways.

Rather, consider the finite Fourier transform with a Hanning weighting function χ_p

$$\phi(\omega) = \frac{1}{2T}\int_{-T}^{T} q(t)e^{i\omega t}\chi_p(t/T)dt \tag{8.8}$$

where the Hanning weighting function is

$$\chi_p(t/T) = \frac{2^p(p!)^2}{(2p)!}\left(1+\cos\left(\frac{\pi t}{T}\right)\right)^p \tag{8.9}$$

Weighting functions are often used in signal processing to speed the convergence of frequency estimates. Figure 8.2 shows a plot of the finite Fourier transform of a single spectral line at frequency ω_0, using different Hanning window functions χ_p. First, notice that although only one spectral line is actually present, each case shows an infinite series of *sidelobes.* These are spurious peaks, and must be rejected. Also, as p increases, the width of the main peak increases, so that it is much wider than the spurious sidelobes. The sidelobe frequency is easily found to be

$$\omega_T = \pi/T \tag{8.10}$$

so their presence is easily recognized. Finally, as p increases, the far field value of the transform drops more dramatically, suppressing the sidelobes in the process. The use of a Hanning window function also increases the convergence rate in seeking the exact frequencies. Laskar has shown that for $p = 1$ that accuracy of frequencies converges as $1/T^4$. This makes it possible to determine the system frequencies with exquisite accuracy.

In order to first identify the basis frequencies, it is necessary to search for maxima in the magnitude of (8.8) near an approximately known frequency ω. The power spectrum is what is commonly plotted when the Fourier transform is displayed, and it is given by $P = |\phi|^2$. To set up a Newton-Rhapson algorithm to find the maxima, we need to find the zero of the slope $d|\phi|^2/d\omega$, which in the vicinity of a peak can be expanded as

$$\frac{\partial P}{\partial \omega} = 0 \approx \left.\frac{\partial P}{\partial \omega}\right|_{\omega_0} + \left.\frac{\partial^2 P}{\partial \omega^2}\right|_{\omega_0} \delta\omega \tag{8.11}$$

about a reference frequency ω_0. This supplies the correction to this frequency, $\delta\omega$ if the first and second derivatives can be calculated. Simple calculation gives the differential equations for these frequency derivatives as

$$\frac{d}{dt}\phi = \frac{1}{2T}q(t)e^{i\omega t}\chi_p(t/T) \tag{8.12}$$

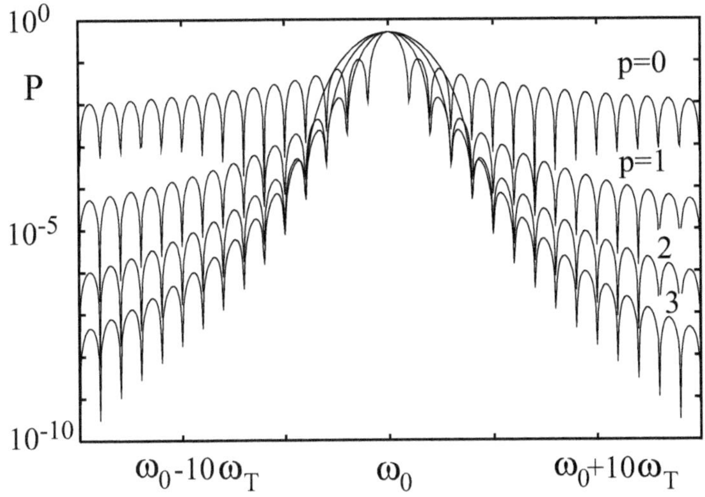

Figure 8.2: The finite Fourier transform of a single input frequency ω_0.

$$\frac{d}{dt}\frac{\partial\phi}{\partial\omega} = \frac{i}{2T}tq(t)e^{i\omega t}\chi_p(t/T) \tag{8.13}$$

$$\frac{d}{dt}\frac{\partial^2\phi}{\partial\omega^2} = -\frac{1}{2T}t^2q(t)e^{i\omega t}\chi_p(t/T) \tag{8.14}$$

with zero initial conditions for all equations. Then, the required derivatives of the power are simply found as

$$\frac{\partial P}{\partial\omega} = 2\Re\phi\frac{\partial\Re\phi}{\partial\omega} + 2\Im\phi\frac{\partial\Im\phi}{\partial\omega} \tag{8.15}$$

$$\frac{\partial^2 P}{\partial\omega^2} = 2\left(\frac{\partial\Re\phi}{\partial\omega}\right)^2 + 2\left(\frac{\partial\Im\phi}{\partial\omega}\right)^2 + 2\Re\phi\frac{\partial^2\Re\phi}{\partial\omega^2} + 2\Im\phi\frac{\partial^2\Im\phi}{\partial\omega^2} \tag{8.16}$$

Here $\Re$ is the real part, and $\Im$ is the imaginary part of the following expression. As multiple passes through the trajectory history $\mathbf{q}(t)$ are usually necessary, it is best to store the results of the numerical integration, and then calculate the finite Fourier transforms outside the orbit integration.

Figure 8.3 shows two power spectra of an orbit with semimajor axis $a = 1.05$ earth radii, an eccentricity of $e = 0.05$, and an inclination of $i = 30$ degrees. The top plot was produced by propagating the orbit using the Simplified General Perturbations 4 model, and transforming the result to the earth centered rotating frame. The lower plot comes from a numerical integration of (8.6) with a 20 by 20 earth gravity model. Time Units (TU) of approximately 13.44 minutes were used, making the usual Keplerian frequency $n \approx 1$. The spectra

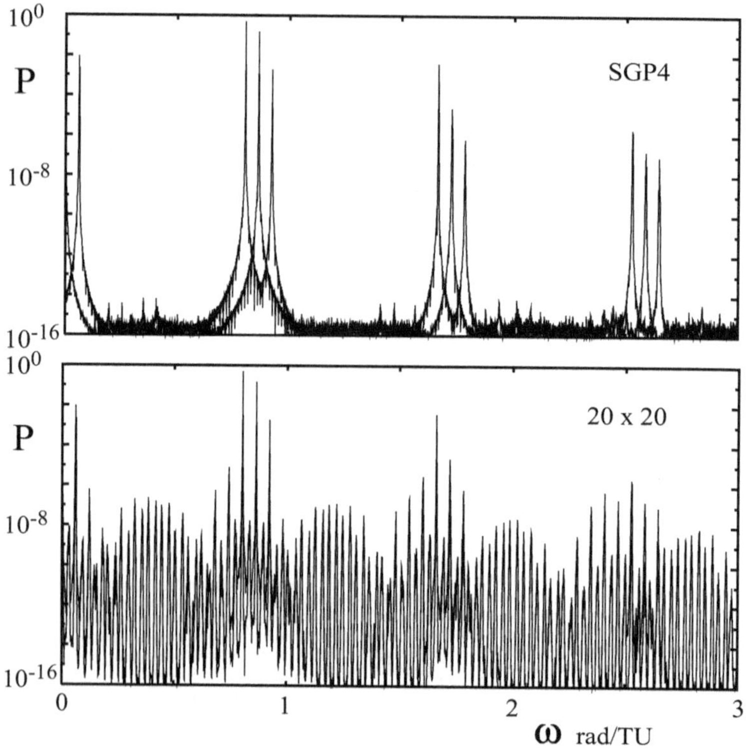

Figure 8.3: Power spectra of an orbit represented by a simplified perturbation theory (SGP4) and a 20 by 20 numerical integration.

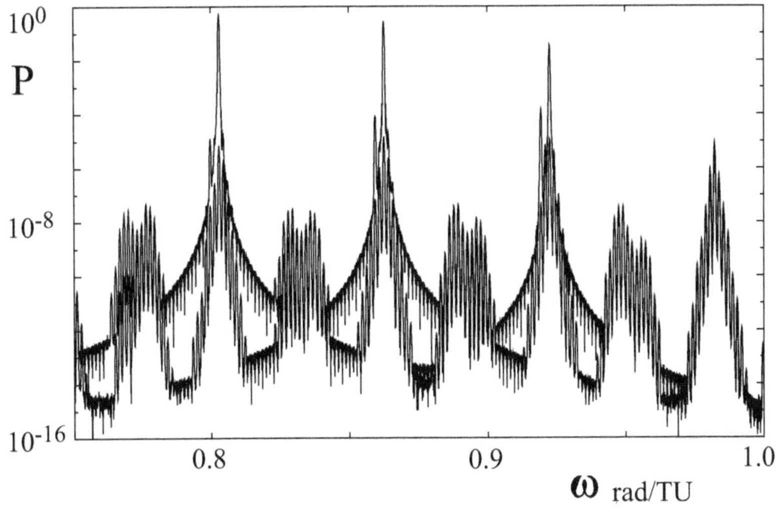

Figure 8.4: Detail of power spectrum in the vicinity of the Keplerian frequency triple.

from the full numerical integration is imposing, but the perturbation theory helps to make sense of this. The simple version shows peaks mainly in groups of three. The center peak is from the z coordinate, and occurs at multiples of the Keplerian frequency. This is flanked by two peaks from the x and y coordinates, which are offset to each side by the earth's rotation rate, plus the nodal rate. Two isolated peaks at lower frequencies are the earth's rotation rate minus the node rate, and very close to zero, the drift rate of the argument of perigee. That is, approximately through order J_2, the main frequencies are

$$\begin{aligned} \omega_1 &\approx \dot{M} = \sqrt{\frac{\mu}{a^3}} - \frac{3\sqrt{\mu}J_2R_\oplus^2}{2a^{7/2}(1-e^2)^{3/2}}\left(\frac{3}{2}\sin^2 i - 1\right) \\ \omega_2 &\approx -\omega_\oplus + \dot{\Omega} = -\omega_\oplus - \frac{3\sqrt{\mu}J_2R_\oplus^2}{2a^{7/2}(1-e^2)^2}\cos i \\ \omega_3 &\approx \dot{\omega}_{TBP} = -\frac{3\sqrt{\mu}J_2R_\oplus^2}{2a^{7/2}(1-e^2)^2}\left(\frac{5}{2}\sin^2 i - 2\right) \end{aligned} \tag{8.17}$$

where the two body problem argument of perigee has been denoted ω_{TBP} to avoid some confusion. These are approximations, however. In a sense, a KAM torus is a *converged* perturbation theory, so the actual basis frequencies need to be established from the spectra, as discussed above.

The previous figure understates the detail visible in the power spectrum. Figure 8.4 shows an enlarged area around the main Keplerian frequency triplet. Multiples of the earth's rotation rate appear as the orbit responds to the details of the gravity field of the earth's continents and ocean basins. This appears as

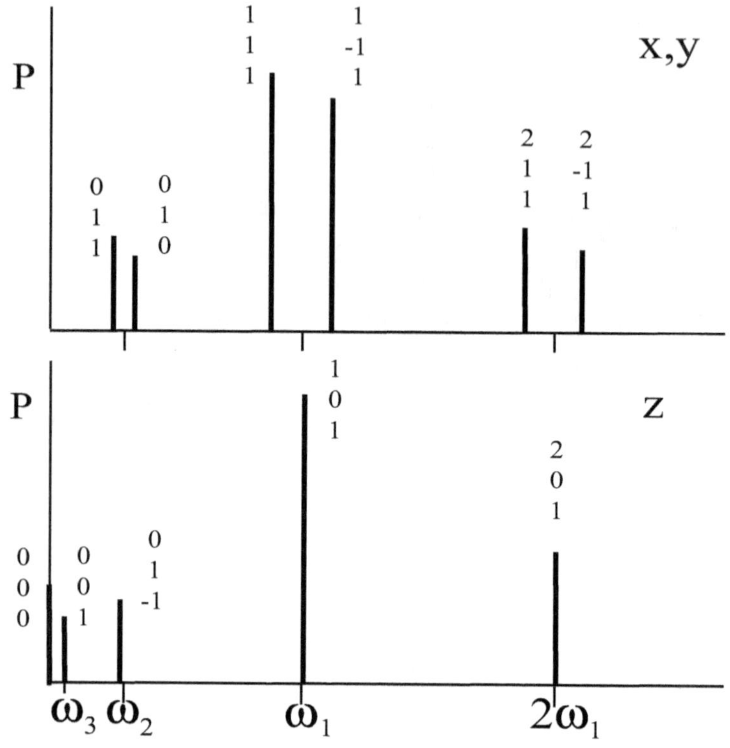

Figure 8.5: Identification of some important spectral lines in terms of their basis frequencies.

clusters of spectral lines separated by the earth's rotation rate. Each cluster shows fine detail that is separated by the argument of perigee rate. This drops off rapidly, since each multiple of the argument of perigee effectively brings in another power of the eccentricity. Note, for reference in the next section, that these clusters of lines are decoupled from each other to high order.

Knowing the sidelobe frequency ω_T also allows spurious peaks to be separated from the actual dynamics signature, with an appropriate choice of integration span T. The sidelobe structure can be glimpsed on the shoulders of the largest peaks in Figure 8.4. It is clearly different from the shortest ω_3 spacing. Also, one hallmark of chaotic systems is that the spectrum becomes *infinitely* detailed. That is not the case here, at least not to double precision accuracy. There are only a finite number of lines in this figure, not a continuous spectrum.

In order to reduce this to a set of basis frequencies, Figure 8.5 shows a schematic version of the earlier SGP4 spectra, with identifications based on their perturbation theory rates. These spectral lines can be found, and their actual frequencies identified by the Newton-Rhapson iteration derived earlier.

Table 8.1: Approximate J_2 and estimated frequencies ω_i.

i	ω_i, J_2	ω_i torus
1	0.861187748359008	0.861159538492850
2	-0.059829329424859	-0.059833012220000
3	0.001580945356222	0.001585676087884

Table 8.2: Coordinate, line identity, observed frequency, and residual from the basis frequency fit.

coordinate	**j**	ω_i	residual
x	1 1 1	0.802912202360808	7.3×10^{-14}
y	1 1 1	0.802912202360810	7.5×10^{-14}
x	1 -1 1	0.922578226800634	-1.0×10^{-13}
y	1 -1 1	0.922578226800576	-1.6×10^{-13}
z	1 0 1	0.862745214580843	1.1×10^{-13}
x	2 1 1	1.66407174085347	-1.2×10^{-13}
y	2 1 1	1.66407174085344	-1.4×10^{-13}
x	2 -1 1	1.78373776529280	-7.8×10^{-13}
y	2 -1 1	1.78373776529452	9.3×10^{-13}
z	2 0 1	1.72390475307370	1.2×10^{-13}

For the orbit mentioned earlier, a simple least squares estimator was used to estimate the three basis frequencies from the observed peak position of these lines. That is, the basis frequencies were fit according to

$$\omega_{observed} = j_1\omega_1 + j_2\omega_2 + j_3\omega_3 \tag{8.18}$$

where the j_i are as shown alongside the lines in Figure 8.5. This gives angle variables Q_i which will resemble their classical counterparts. The resulting basis frequencies are given in Table 8.1, along with the approximate J_2 frequency values. The units are radians per time unit (TU). Once this fit was made, the observed line frequencies could be compared to the estimated linear combinations, and their residuals computed. This leads to Table 8.2, which shows that the observed frequencies from the spectra were fit to double precision accuracy. This is also strong evidence that the spectra peaks are not random, but really do arise as linear combinations of only three basis frequencies.

8.3 Extracting Fourier Coefficients

With a knowledge of the frequency basis firmly in hand, it is not necessary to scan the entire spectrum for spectral lines. Their position can be accurately predicted, still assuming that this is a KAM torus. If we perform the finite Fourier transform on a single spectral line

$$q(t) = \mathcal{C}_i \cos\omega_i t + \mathcal{S}_i \sin\omega_i t \tag{8.19}$$

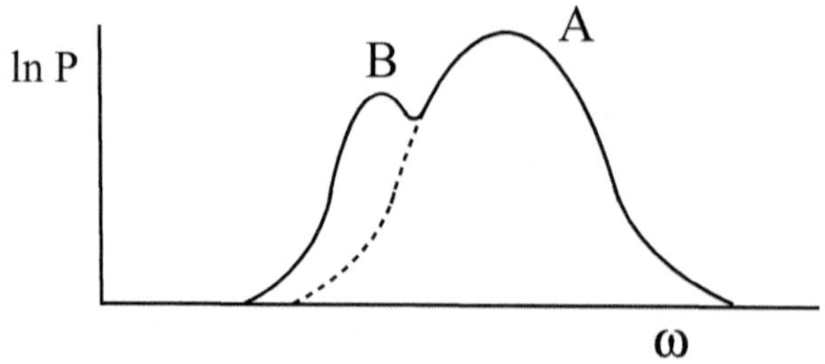

Figure 8.6: Two superimposed spectral lines

and evaluate at the line peak frequency, we obtain

$$\phi(p=0) = \mathcal{C}_i\left(\frac{2\omega_i T + \sin 2\omega_i T}{4\omega_i T}\right) + i\mathcal{S}_i\left(\frac{2\omega_i T - \sin 2\omega_i T}{4\omega_i T}\right) \quad (8.20)$$

$$\begin{aligned}\phi(p=1) &= \mathcal{C}_i\left(\frac{2\omega_i T(\pi^2 - 4\omega_i^2T^2) + \pi^2 \sin 2\omega_i T}{4\omega_i T(\pi^2 - 4\omega_i^2 T^2)}\right) \\ &+ i\mathcal{S}_i\left(\frac{2\omega_i T(\pi^2 - 4\omega_i^2T^2) - \pi^2 \sin 2\omega_i T}{4\omega_i T(\pi^2 - 4\omega_i^2 T^2)}\right)\end{aligned} \quad (8.21)$$

$$\begin{aligned}\phi(p=2) &= \mathcal{C}_i\left(\frac{2\omega_i T(\pi^4 - 5\pi^2\omega_i^2T^2 + 4\omega_i^4T^4) + \pi^4 \sin 2\omega_i T}{4\omega_i T(\pi^4 - 5\pi^2\omega_i^2T^2 + 4\omega_i^4T^4)}\right) \\ &+ i\mathcal{S}_i\left(\frac{2\omega_i T(\pi^4 - 5\pi^2\omega_i^2T^2 + 4\omega_i^4T^4) - \pi^4 \sin 2\omega_i T}{4\omega_i T(\pi^4 - 5\pi^2\omega_i^2T^2 + 4\omega_i^4T^4)}\right)\end{aligned} \quad (8.22)$$

So, evaluating the Fourier transform at the known line center frequency allows us to easily extract the sine $\mathcal{S}_i$ and cosine $\mathcal{C}_i$ coefficients.

However, another problem presents itself, shown schematically in Figure 8.6. The larger spectral peak A underlies the smaller peak B, and contributes substantially to its apparent amplitude. This also happens in the reverse sense, and just as importantly, since the coefficient of a larger amplitude line must be known to correspondingly more significant figures. This problem can be mitigated somewhat by using higher order Hanning weighting functions, since these fall off more rapidly with distance from the central peak. But since the usual spectral power plot is logarithmic, and the Fourier coefficients may span many orders of magnitude, this cannot be neglected.

The solution proposed by Laskar is to process the entire spectrum to find the largest amplitude spectral line, and to fit its amplitude. Then, knowing the sine and cosine coefficients for this line, its contribution can be subtracted from the orbit data

$$\mathbf{q}'(t_i) = \mathbf{q}(t_i) - \mathcal{C}\cos\omega t_i - \mathcal{S}\sin\omega t_i \quad (8.23)$$

With the largest line removed, the spectrum can be recomputed, and the next largest line removed from the spectrum. This process may need to be iterated more than once, since even the largest amplitude line may lay over the sidelobe of a smaller line. This may be preferable to trying to solve a huge set of linear equations for all of the Fourier coefficients simultaneously. On the other hand, there is another option in the case of low eccentricity earth satellites.

Figure 8.3 on the top shows the spacing of the Keplerian frequency ω_1. The lower half of the figure shows the inclusion of many more peaks, now separated by the earth rotation / nodal regression frequency ω_2. Finally, Figure 8.4 shows that each of these peaks is really a cluster of spectral lines separated by the smallest frequency, the argument of perigee frequency ω_3. Each of these line clusters is almost perfectly decoupled from the adjacent cluster, since $\omega_2 >> \omega_3$. Even if two clusters should overlap, they would still only influence each other, and would be decoupled from adjacent clusters. So it is possible to solve a very small order set of linear equations for the coefficients $\mathcal{C}_i$, $\mathcal{S}_i$, instead of solving a huge linear system, or iteratively subtracting one line at a time. Since solving a linear system scales on the order of N^3, where N is the number of unknowns, a moderate number of small order systems is much to be preferred over one large order system. This method has recently been studied by Bordner and the current author.

This torus construction algorithms simply places the origin $Q_i = 0$ at the epoch time of the integration. Since these angles are linear in time, we are free to adjust their origins Q_{i0} to more nearly align with their classical definitions. Given the torus Fourier series, this can be done as a three step process. First, holding Q_2 and Q_3 at zero, the value of Q_1 can be adjusted to minimize the radius vector in the range $0 \leq Q_1 \leq 2\pi$. This places the origin of Q_1, the mean - anomaly analogue, at perigee. Then, holding Q_1 at perigee, and holding the longitude of the node Q_2 constant, the value of Q_3 is found that places the satellite on the equator crossing from south to north. This sets the argument of perigee analogue's origin at the ascending node. Finally, with the origins of Q_1 and Q_3 identified, the origin of the longitude of the ascending node is adjusted to place the node on the Greenwich meridian. This sets the origin of the torus coordinates Q_i at the closest analogue values to their classical two body element counterparts. Since both approaches are describing the same problem, it is not surprising that there is a close correspondence between them.

There is one major flaw with the method outlined here for extracting a KAM torus, and that is the dependence on a long numerical integration of the orbit. The example shown here was based on a one year integration. This fails when one or more of the basis frequencies become small. Then the usual Nyquist criteria of signal processing mandates a very, very long numerical integration to try to identify the small frequency. This happens near the critical inclination, and it also happens in the vicinity of an orbital resonance.

This does not mean that KAM theory cannot be very successful away from resonances and areas where small frequencies require long integrations. Figure 8.7 shows the example orbit, fit to a one year integration, now compared to a 10 year numerical integration. Over this interval, the orbit makes approximately 400,000 revolutions, so in predicting the satellite's position from the

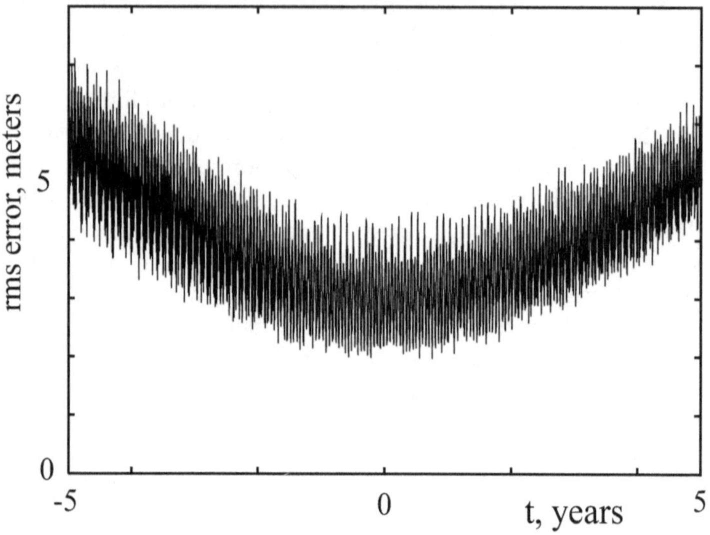

Figure 8.7: Root mean square residuals for a ten year numerical integration compared to a one year KAM torus fit

Fourier series, more than double precision arithmetic is called for in calculating the products $\omega_i t$, before reducing them modulo 2π. Otherwise the "error" is dominated by truncation in these terms. The figure plots a sliding root mean square distance, in meters, between a numerical integration and the KAM torus, again using a 20 by 20 gravity field. The starting time is in the middle, so an error growth of 5 meters over 5 years, assuming it is an in-track error, implies a frequency error of about $\delta\omega_1 \approx 4 \times 10^{-12}$ radians /TU. This is slightly more pessimistic than the basis frequency estimates of Table 8.2, and could be due, of course, to the numerical integration.

The author is under no delusion that an actual orbit could be predicted to this accuracy for this interval of time. But it is astonishing confirmation that the KAM torus *is* the solution to the gravitational problem of orbiting a nonspherical earth.

8.4 Motion Near A KAM Torus

There is one major obstacle to using a KAM torus to model an earth satellite orbit. As currently posed, the torus itself has explicit dependence on three new coordinates Q_i, but while the Poincaré invariants make it possible to calculate the new conjugate momenta P_i, this is true only on the surface of the torus itself. In order to be useful, small corrections may need to be made in all six variables, and this means that some description of motion *near* a KAM torus is needed. While tori may not be continuous throughout the nearby phase space, they will probably be dense. This means that most nearby trajectories will also lie on KAM tori.

In a very real sense, this is a realization of the Hamilton-Jacobi theorem. The new momenta P_i are all constants of the motion, while the new coordinates Q_i increment linearly with time. A dynamical system with this property must come from a Hamiltonian $\mathcal{K}$ that is a function of the new momenta only, and is free of the new coordinates. It is not too difficult to approximate such a Hamiltonian. The Delaunay momenta are close approximations to the actual torus P_i, and the frequencies $\omega_i = \dot{Q}_i$ are closely approximated by the well known secular rates. While we do not wish to give up using the actual momenta and frequencies of the torus, the linearization might be approximated by a Hamiltonian

$$\mathcal{K} = -\frac{\mu^2}{2P_1^2} - \omega_{\oplus} P_2 + \frac{\mu^4 J_2 R_{\oplus}^2 (P_3^2 - 3P_2^2)}{4P_1^3 P_3^5} \tag{8.24}$$

The first term is the two body term in Delaunay elements, the second term arises from using a reference frame that rotates with the earth. The third term is the secular J_2 potential transformed into the Delaunay variables. This form is sufficient to argue the structure of nearby trajectories.

Let the state vector $\mathbf{X}^T = (\mathbf{q}^T, \mathbf{p}^T)$ be the physical variables, and correspondingly $\mathbf{Y}^T = (\mathbf{Q}^T, \mathbf{P}^T)$ be the torus canonical state vector. This means that the Jacobian $\partial\mathbf{X}/\partial\mathbf{Y}$ must be a symplectic matrix, as in section 2.12. Expanding for small deviations from a reference torus, we have

$$\delta\mathbf{X}(t) = \frac{\partial\mathbf{X}}{\partial\mathbf{Y}}\delta\mathbf{Y}(t) \tag{8.25}$$

where the Jacobian must be evaluated on the reference torus. But we can easily argue the form of $\delta\mathbf{Y}(t)$. The last three components of $\delta\mathbf{Y}$ are the δP_i, and these must be constant on an adjacent torus. Expanding the coordinate equations of motion

$$\dot{Q}_i = \frac{\partial\mathcal{K}}{\partial P_i} = \omega_i(\mathbf{P}) \tag{8.26}$$

about the reference torus gives

$$\begin{aligned} \dot{Q}_i &= \omega_i + \frac{\partial\vec{\omega}}{\partial\mathbf{P}}\delta\mathbf{P} \\ &= \omega_i + \frac{\partial^2\mathcal{K}}{\partial\mathbf{P}^2}\delta\mathbf{P} \end{aligned} \tag{8.27}$$

An approximation to this symmetric matrix

$$\frac{\partial^2\mathcal{K}}{\partial\mathbf{P}^2} = \frac{\partial\vec{\omega}}{\partial\mathbf{P}} \tag{8.28}$$

can easily be calculated from (8.24). Also the matrix $\partial\mathbf{X}/\partial\mathbf{Y}$ can also be approximated by (laboriously) calculating the partial derivatives of the transformation from physical coordinates to classical elements, and then from classical

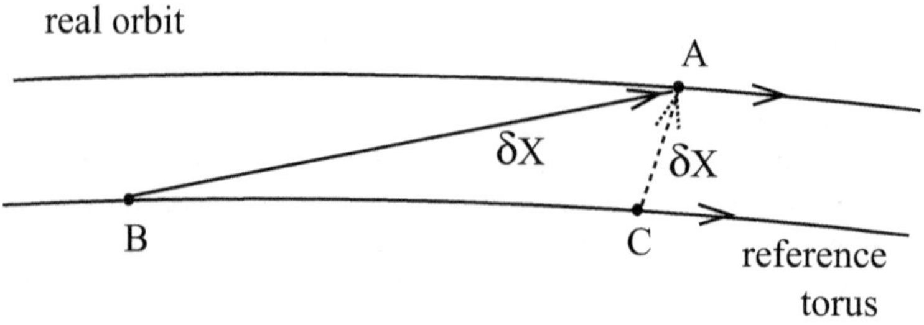

Figure 8.8: Implementing coordinate δQ drift within the reference torus keeps $\delta \mathbf{X}$ small.

elements to Delaunay coordinates. This is only an approximation, but it does serve as an initial approximation.

This is only a linearization of motion around a KAM torus. It is necessary to keep $\delta\mathbf{X}$ small for this approach to be valid. The displacements $\delta\mathbf{P}$ will hopefully remain small, or a torus will not be a good reference trajectory for very long. However, linear drift in the coordinates $\delta\mathbf{Q}$ is inevitable, and then the true position at point A in Figure 8.8 will depart further and further from the reference torus position, point B. To keep this from happening, it is simple to calculate the reference torus coordinates $\mathbf{Q}$ by integrating (8.26) instead of simply using the zeroth order approximation $\mathbf{Q} = \mathbf{Q}_0 + \vec{\omega}t$. This will keep the reference torus much closer (point C) to reality, and hopefully extend the range of the approximation. Of course, $\partial\mathbf{X}/\partial\mathbf{Y}$ should be evaluated at the current torus position, point C.

This technique was used to produce Figure 8.9, where initial displacements from the reference torus were chosen to be δP_1 small, and all other displacements were zero. As expected, this induces an in-track drift in Q_1, and the other coordinates drift much more slowly. A similar plot would show that the δP_i are virtually constant at their initial values. The deviation from the predicted straight line behavior arises from the use of the two body problem / J_2 classical solution to obtain the linearization for the KAM torus. Obtaining this linearization directly from the torus itself, to higher accuracy, is a topic of current research.

Finally, notice that the ability to produce equations of motion for the displacement from the reference torus arises from the expansion of the Hamiltonian. This is a linear system with its own Hamiltonian

$$\begin{aligned}\delta\mathcal{K} &= \frac{1}{2}\delta\mathbf{Y}^T\frac{\partial^2\mathcal{K}}{\partial\mathbf{Y}^2}\delta\mathbf{Y} \\ &= \frac{1}{2}\left(\delta\mathbf{Q}^T, \delta\mathbf{P}^T\right)\left\{\begin{array}{cc} \phi & \phi \\ \phi & \partial\vec{\omega}/\partial\mathbf{P}\end{array}\right\}\left(\begin{array}{c}\delta\mathbf{Q} \\ \delta\mathbf{P}\end{array}\right) \qquad (8.29)\end{aligned}$$

where ϕ represents a three by three zero matrix, and $\partial\vec{\omega}/\partial\mathbf{P}$ is a three by

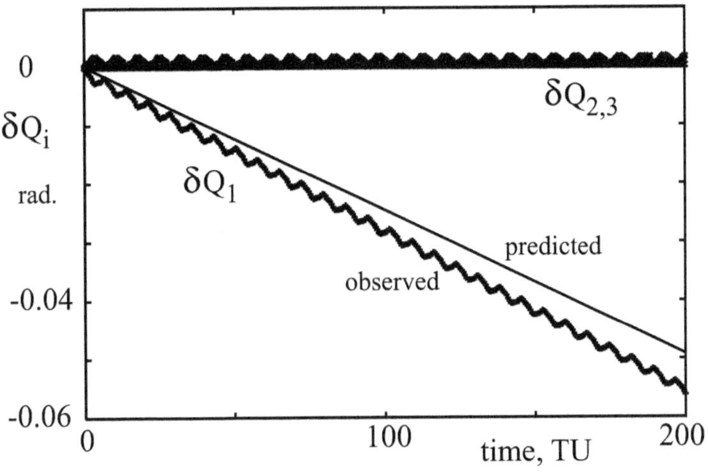

Figure 8.9: Nearly linear drift in δQ_1 when δP_1 is displaced.

three constant matrix when evaluated on the torus. So this is a constant coefficient linear system. Since it is Hamiltonian, it is quite simple to contemplate adding perturbing potentials to the linearized Hamiltonian. Alternately, for non-conservative forces, the physical perturbing acceleration $\mathbf{a}_p$ would be expressed in the components of the earth centered rotating frame, to be compatible with (8.6), where the momenta $\mathbf{p}$ are the inertial velocity components resolved in this frame. Then, this can be transformed to the torus variables $\mathbf{Q}$, $\mathbf{P}$ to give perturbation equations of motion

$$\begin{pmatrix} \delta\dot{\mathbf{Q}} \\ \delta\dot{\mathbf{P}} \end{pmatrix} = \left\{ \begin{array}{cc} \phi & \partial\vec{\omega}/\partial\mathbf{P} \\ \phi & \phi \end{array} \right\} \begin{pmatrix} \delta\mathbf{Q} \\ \delta\mathbf{P} \end{pmatrix} + \frac{\partial\mathbf{Y}}{\partial\mathbf{X}} \begin{pmatrix} 0 \\ \mathbf{a}_p \end{pmatrix} \tag{8.30}$$

since the kinematic terms involving $d/dt(\partial\mathbf{Y}/\partial\mathbf{X})$ have already been absorbed in writing (8.29). So, it is possible to do perturbation theory starting from a KAM torus as the reference solution.

8.5 Discussion and References

This volume has spanned the sweep of orbital mechanics techniques, from the literal variable expansions of the seventeenth century, through Hamiltonian mechanics to the current research topic of KAM tori. While some of these methods have been updated [for example: the ability to do literal mathematics with such programs as Mathematica, Maple, or Matlab] each technique is in a way suited to the technology of its own time. Even though programming literal series expansions is possible, it is computationally inefficient compared to, e.g. numerical integration.

KAM tori holds the promise of updating the old technique of general perturbations to an accuracy that competes with numerical integration. With its heavy dependence on digital computation, numerical integration and KAM tori are suited to an age that promises to be increasingly rich in computational resources. The dependence of KAM theory on a long numerical integration is one area that needs to be investigated. It should be possible, for example, to calculate a KAM torus directly from the equations of motion, without having to go through a numerical integration. The field is progressing rapidly, and this is only a sketchy introduction. The author would like to reference the seminal papers of Laskar:

Laskar, J.: "Frequency Map Analysis and Quasiperiodic Decompositions", Proceedings of Proquerolles School, 2001, 1-31.
Laskar, J.: "Introduction to frequency map analysis", in *Hamiltonian Systems with Three or More Degrees of Freedom,*, Simo, C. (ed.), Kluwer Academic, Netherlands, 1999, 134-150.

Any reader who seriously wishes to begin working with KAM tori should, of course, search the current literature.

Index

www.ingramcontent.com/pod-product-compliance
Lightning Source LLC
Chambersburg PA
CBHW071416170526
45165CB00001B/300
* 9 7 8 1 4 5 3 7 8 1 4 7 0 *